Slaveiko Gospodinov

Space astronomy

Slaveiko Gospodinov

Space astronomy

ScienciaScripts

Imprint

Any brand names and product names mentioned in this book are subject to trademark, brand or patent protection and are trademarks or registered trademarks of their respective holders. The use of brand names, product names, common names, trade names, product descriptions etc. even without a particular marking in this work is in no way to be construed to mean that such names may be regarded as unrestricted in respect of trademark and brand protection legislation and could thus be used by anyone.

Cover image: www.ingimage.com

This book is a translation from the original published under ISBN 978-620-6-78487-6.

Publisher:
Sciencia Scripts
is a trademark of
Dodo Books Indian Ocean Ltd. and OmniScriptum S.R.L publishing group

120 High Road, East Finchley, London, N2 9ED, United Kingdom
Str. Armeneasca 28/1, office 1, Chisinau MD-2012, Republic of Moldova, Europe
Printed at: see last page
ISBN: 978-620-6-59338-6

Contents

Introduction

Space astronomy emerged about 30 years after the launch of the first artificial earth satellites (AES). Since the launch of the AES, Earth remote sensing (ERS) has developed intensively. remote sensing has a technological direction of research from space to Earth. Space astronomy [1] is directed from artificial earth satellites and spacecraft into space, i.e. in the direction back to Earth. Figure 1 shows the research direction of astronomy, space astronomy and Earth remote sensing (ERS) technologies.

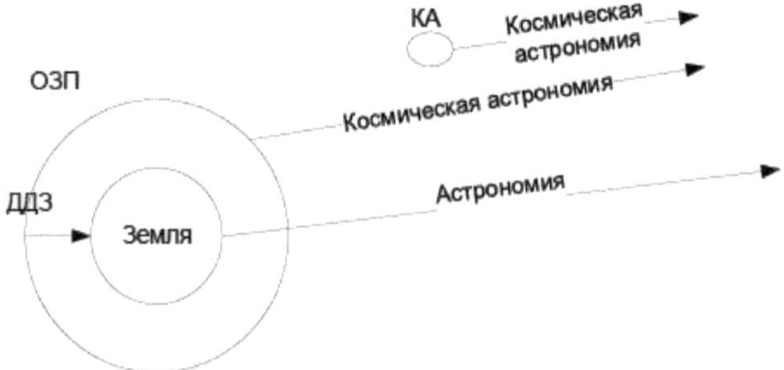

Figure 1. Research focus of astronomy, space astronomy and Earth remote sensing technologies.

In Figure 1, the following designations are introduced: OZP - near-Earth space, RS - remote sensing of the Earth, spacecraft - spacecraft.

Classical astronomy is driven by the needs of human practice for terrestrial and stellar coordination, for terrestrial navigation, for determining synchronous time, and for studying the chronology of development.

Classical astronomy studies the patterns of stellar and planetary arrangements. Astronomy studies the motion of celestial bodies, the structure of planetary systems, and the development of celestial systems and individual stars. Many definitions of astronomy do not mention the focus of the study of astronomy. Astronomical research has long been done from the surface of the Earth into space. The development of the AES, orbital stations, and spacecraft made it possible to move observations from the Earth's surface into outer space. This has created two directions for such observations. The first direction of observations is from space to Earth. It is called remote sensing of the Earth. The second direction of observations from space to space was not called for a long time. Recently the name "space astronomy" has appeared for it. Space astronomy differs from classical astronomy not only in the transfer of observation points from the Earth's surface to outer space, but also in its methodological and organisational basis. In some cases, space astronomy is realised as a technological science. Its methods are aimed at solving applied problems. Space astronomy can be compared to geoinformatics, which at the beginning of its development was created as a technological field.

The organisational components of space astronomy are: classical astronomy, geodesic astronomy [2, 3], space geoinformatics [4-7], space photogrammetry, dynamic photogrammetry, dynamic geoinformatics [8], spatial logic [9-11], systems analysis [12], computer science [13, 14] and space geodesy [15, 16]. Classical astronomy developed independently of geoinformatics. The main means of collecting information in space astronomy is space monitoring [17-19], which is conducted in different spectral bands. Another difference between classical astronomy and space astronomy is the types of observational objects. Classical astronomy is an "object" and "system" science. It studies objects and their systems. Space astronomy is an "object", "system" and "situation" science. It investigates spatial information situations in which objects are located, as well as systems of objects and systems of situations. Space astronomy considers the world as a system of systems [20] Space astronomy considers outer space as a system of nested spaces [18]. In the modern world, information serves as the basis of development. Therefore, space astronomy uses methods of informatics, which did not exist at the time of the emergence and application of conventional astronomy.

Modern computer science uses the concept and model of information field [21-23] for integrated processing of observations. The concept of the information field of information is applied in space research [22]. It is logical to apply this concept in space astronomy. Thus, space astronomy is a complex of sciences and its study is relevant.

Cosmic spatial logic is an important component of space astronomy. Logic is interpreted as the science of reasoning. More recently, logic has been interpreted as the science of finding and describing patterns of the world around us. For space astronomy, spatial logic is of significance. Spatial logic [9, 10, 11, 24] describes the regularities of spatial systems and structures. Spatial logic in space astronomy includes geometric logic, image logic and visibility logic. The logic of visibility is due to the peculiarity of space observations. Objects such as planets are large in size compared to terrestrial objects. Space objects move at enormous speeds.

Spatial logic is used in trajectory calculations and analyses. Observation distances in space are orders of magnitude greater than those in terrestrial conditions. Spatial logic is used in building virtual models of outer space. Logic is needed in the study of near-Earth space and the study of planetary surfaces.

A special feature of space astronomy and space geoinformatics is the extensive use of angular measurements [25, 26]. Many studies of planets do not have the possibility of linear contact direct measurements. Most measurements in astronomy and space astronomy are angular. Cosmic spatial logic allows for spatial and geometric constructions. It relates real angular measurements to linear dimensions of celestial bodies.

Coordinate systems in space astronomy are applied and developed on the basis of astronomical coordinate systems and terrestrial coordinate systems. Many tasks of space astronomy are related to measurements on the surface of planets. Different coordinate systems are used on the surfaces of planets. the principle of systems selection is similar to terrestrial coordinate systems [27]. For modelling the surface of planets, planetocentric, external, reference (model), surface (topocentric) coordinate systems and fragmentary (local) systems are used. On planets, geodetic networks are generally (with the exception of the Earth) difficult to establish. Planetocentric coordinate systems are related to the planet's centre of mass and are most often spherical. External coordinate systems have a conventional centre. They are related to an external point in space and are not related to the planet. They are usually also spherical or elliptical. Such a system is created in a notional space relative to the planet and requires points with known coordinates on the planet's surface to anchor it to the surface. GPS, GLONASS systems are examples of such systems. The planet is nested in this coordinate space and there is a need to link this space with the surface or centre of mass of the planet.

Surface (topocentric with respect to the Earth's surface) coordinate systems are created with respect to surface points. They are referenced to the surface points. As a rule, they are Cartesian coordinate systems. These systems are referenced to the planetocentric system. Local coordinate systems are created in small areas to solve engineering problems. They are floating and have no reference to the planetocentric coordinate system. They are analogues of surface coordinate systems. They are created for the duration of engineering or construction works.

Planetocentric coordinate systems are associated with the planet's centre of mass. These coordinate systems are based on a model of the planet: either a spheroid or an ellipsoid. For the ellipsoid model, the parameters of the ellipsoid, the equatorial plane, the poles, and the centre must be specified. Planetary ellipsoid serves for modelling the shape of a planet, if it is similar to an ellipsoid. Therefore, the main task of such a model is to simulate the shape.

Reference coordinate systems are associated with the reference ellipsoid model. These systems are used for the studied areas. An additional procedure is the orientation of the reference ellipsoid in the body of the planet. Reference ellipsoids define a system of altitudes. Therefore, in contrast to the general planetary ellipsoid for orientation and fixing of the reference ellipsoid in the planet body it is necessary to set initial geodetic dates (*datum*). The main task of such a model is modelling of heights on the planet surface.

The topocentric coordinate system is rectangular. Its
the origin is on the planet surface, near the surface or under the surface. The direction of the Z axis is normal to the surface of the planet model (spheroid or ellipsoid) The X axis lies in the plane of the meridian that passes through the origin. It is directed towards the north pole. The Y axis completes the formed system to the left. The system participates in the daily rotation of the planet, remaining stationary relative to the points of the surface.   Such conditions of the assignment
The planetocentric and topocentric systems ensure comparability of measurements on different planets.

The Z-axis of the topocentric system defines the conventional vertical and serves as the basis for measuring heights on the planet. However, there are many contradictions in the geodetic literature. "A vertical is a straight line in space perpendicular to a horizontal plane". Horizontal planes do not exist in nature. The surface of any planet is not horizontal but a convex surface. Most major planets have a shape close to spherical.

Because of this, other approaches must be used to determine heights.

Many planets have the geometrical centre of the planet figure displaced relative to the planet's centre of mass. Because of this, the plumb line direction does not correspond to the vertical direction for the reference ellipsoid.

Relation of space astronomy to computer science and geoinformatics.

Informatics [14] and geoinformatics [28] have had a great impact on the development of science and technology. This is mainly due to the use of various models that generalise the properties of real world objects. In addition, informatics and geoinformatics allow the transfer of processing and analysis methods from one domain to another. Space astronomy is closer to applied informatics than to computer science. Space astronomy is closer to applied informatics than to computer science.

Space astronomy is closer to applied geoinformatics [29 than to general geoinformatics. The application of geoinformatics methods is conditioned by the property of its integration [28]. It is expedient to use this property in space astronomy.

Geoinformatics applies artificial intelligence methods [30]. Space astronomy also requires the application of artificial intelligence, especially due to the fact that there is a problem of big data in it. The basis of processing in computer science and geoinformatics is information models [31] and numerical models [32]. This principle is transferred to space astronomy

Space astronomy relies on qualitative and comparative analyses. Comparative analysis in space astronomy uses oppositional analysis [33]. The basis of observation in space geoinformatics is space monitoring. It uses the principles of geomonitoring and geoinformatics monitoring [34]. In space astronomy it is the main source of information acquisition.

In computer science and geoinformatics, digital models are widely used at the intermediate stage of processing. Digital maps are used as a result of processing [32]. In space astronomy, maps are used less frequently and synthesised images and photograms are used more often. Meta-modelling is used to generalise experience in computer science and other sciences [35]. In space astronomy, metamodelling is also used to generalise experience and build hypotheses.

The field concept in space astronomy

The field concept in space astronomy is that the information field model is taken into account in modelling and analysis [22]. Reality is considered by reflecting it into the information field. The information field fulfils three functions [36]: reflection, integration and representation. The reflection mechanism is to reflect reality into the global information model of the information field. The function of integration is that the information field unites disparate and related objects. including their connections and relations. The function of representation is that the information field represents objects and situations in a visual form. It was noted above that space astronomy can be compared to a space photograph. The information field also reflects reality like a photograph. Therefore, conceptually the information field fits into the theory of space astronomy.

The information field includes not only concrete objects but also information uncertainty. Information uncertainty includes information that can be explained logically and on the basis of known theories. According to this feature, explanatory information and uncertainty are distinguished. There is a lot of redundant information in the information field. For example, a bitmap photo contains as much information as a map of an area. But it has an information volume 3-4 orders of magnitude larger than the information volume of a map.

The complement of the information field is the information space. An example of information space in spacecraft is coordinate space. Information space contains different information fields. Information field has a meaningful characteristic of space field function, Field function shows the value of the field characteristic at a given point of space. For example, a satellite receiver shows three-dimensional coordinates at a point in space. The presence of a field function indicates the presence of a field, the absence of a field function indicates the absence of a field. Information space is polymorphic. For example, the coordinate of a point of space can be defined in polar coordinate system, Cartesian coordinate system, spherical coordinate system, cylindrical coordinate system.

Space astronomy emerged as a development of astronomy to solve applied problems. As such, the basis of space astronomy is space geoinformatics. Space geoinformatics developed on the basis of applied geoinformatics, which is oriented towards solving applied problems. Space astronomy, based on the tradition of geoinformatics, is developing by integrating different scientific directions. The first feature of space astronomy integration is its integration with remote sensing technologies. The second feature of space astronomy integration is the transformation of Earth sciences into space disciplines: space geoinformatics, space geodesy, geodesic astronomy and so on. As the third feature of space astronomy we should mention the problem of big data [37-40], which is characteristic of space research. Space astronomy forms spatial knowledge [41] and space knowledge [42}. Space astronomy research methods are aimed at the study of celestial bodies. Space astronomy [43] complements astronomy and space geoinformatics and contributes to the formation of a scientific picture of the world.

1. Space astronomy in the system of sciences

1.1.   Informatics, geoinformatics and space astronomy

Informatics emerged as a science of programming (computer science) and information processing in the context of globalisation and informatisation [44]. The term "informatics" emerged in the 1960s in France to name the field dealing with automated processing of information as a merger of the French words information and automatique (F. Dreyfus, 1962). In different countries, informatics (cf. *German.* Informatik, *English* Information technology, *French* Informatique, *English* computer science - computer science - in the USA, *English* computing science - computing science - in Great Britain) - was interpreted as the science of methods of obtaining, accumulating, storing, transforming, transmitting, protecting and using information. The term "information" was not associated with any subject area, but was a generalisation of information as an *object of* computer *processing*.

Informatics emerged not as "information science" or as "information theory". but as "science of information processing". The main thing in this science was not the study of the content of the concept of "information", but the development of data and information processing methods.

For quite a long time the term "informatics" was used as a synonym for the term "programming". In Russia, computer science courses until 2000 included the study of programming in the first place and the application of information systems in the second place. The object of study (field of research) of informatics or its dominant feature are *methods of information processing*, computer models, algorithms of analysis and calculations - *without regard to* the field of use of this information. It should be emphasised that informatics is concerned with information processing, not information theory. From this point of view, computer science is an *intermediary* between computational mathematics and logic on the one hand and applied sciences on the other. It emerged and develops as a science of information processing.

The application of informatics in any domain is based on the adaptation of informatics to the methods, tasks and data of that subject area. As a result of such specialisation, specific informatics in this subject area appears. For example, economic informatics (business informatics); informatics in biology (bioinformatics); informatics in medicine, informatics in geodesy, informatics in geology, etc. We have not included geoinformatics in this list not by chance. It has a different origin. These "specialised informatics" arises on the basis of *differentiation of informatics in* relation to the problem domain (Fig.1.1). In real practice, the vector of specialised informatics changes significantly.

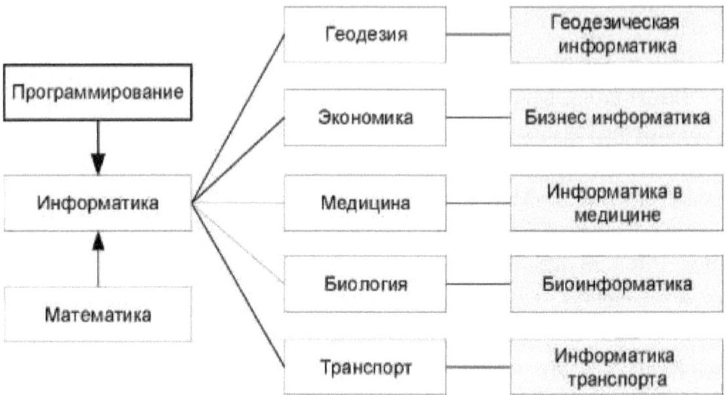

Fig.1.1 Differentiation of informatics

The main thing for specialised informaticians is not information processing, but *solving problems and tasks in their field* using informatics. At this stage it should be noted that any specialised informatics is first of all problem *solving of the subject area* and lastly only information processing or informatics.

In this aspect, we can talk about the application of informatics to space research or space geoinformatics as a set of algorithms. methods and technologies for information processing.

Creation and development of geoinformatics. Unlike informatics, which has one source of origin, geoinformatics emerged from different sources of Earth science (geo); methods of computer-aided design, methods of information processing. The objective need for integration of Earth sciences: geodesy, photogrammetry, cartography, remote sensing of the Earth - has been brewing for a long time, independently of informatics. Teaching of these disciplines was characterised by great interpenetration. In practice, often geodesists worked in photogrammetry, photogrammetrists worked in geodesy or were engaged in map-making, etc.

Space geoinformatics also emerged from the integration of different sciences related to the study of outer space. Informatics was a link, but not the main link, facilitating the integration of these sciences into a single system.

Terrestrial geoinformatics emerged from the integration of Earth sciences and information processing methods into a single system. This integrated system of Earth sciences [45] is called geoinformatics.

Let us emphasise the first important distinction between the application of informatics in other subject areas and in the Earth sciences. In many fields informatics has been applied as a tool for *specialisation of* information processing of a given subject area,

In the earth sciences, computer science is also applied as a specialisation

(informatics in geodesy) and as one of the components of science integration. However, the integrating functions are performed by geoinformatics, which has its own research and processing methods.

Along with geoinformatics there are, for example, informatics in geodesy (until recently such specialists were trained in Russian universities), informatics in geology, informatics in photogrammetry, informatics in cartography. But unlike geoinformatics, these are differentiated sciences that were created on the basis of differentiation, while geoinformatics (an unfortunate name) was created on the basis of integration (Fig. 1.2). It should be noted that foreign synonyms of geoinformatics have the following meanings: "geo science", "geoinformatica" and "geo computer science".

Thus, geoinformatics unlike specialised "informatics" is a system of sciences [45] and is applied in different subject areas. But if informatics carries out interdisciplinary transfer of information processing methods, geoinformatics carries out interdisciplinary transfer of knowledge.

Unlike computer science, geoinformatics has its applied field of study and its specific data. These data are called geodata [46-49].

The data that informatics uses in different fields may not be comparable, they may have different structures, different types and different formats. Data used by geoinformatics in different fields are comparable, they always have the same structure and the same types. The comparability of geoinformatics-geodata makes it attractive for use in space research.

Geodata defines the research area of geoinformatics. This concept is associated with a number of sciences, in which "geo" is formally and meaningfully included as a constituent part (geometry, geodesy, geography, geology, geodynamics, geoinformatics, geomatics, geomarketing, etc.). In Fig. 1.2, thickened lines highlight important factors: earth sciences, geoinformatics and geodata.

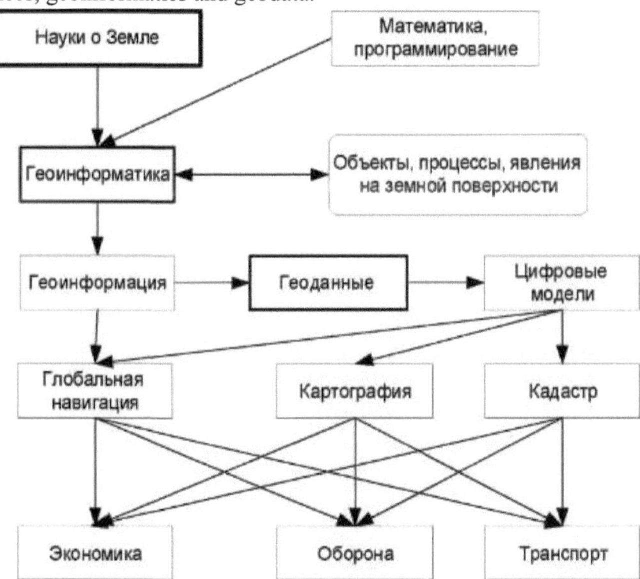

Fig.1.2 Structure and application of terrestrial geoinformatics

The concept of geodata is associated with a number of sciences, which do not explicitly include "geo", but are included in a meaningful way (transport, architecture, earth science, land use, cadastre, real estate management, distributed systems, logistics, space research, photogrammetry, cartography, world economy, social processes and phenomena, human society development, etc.). Thus, the areas covered by the content part of "geo" leads to the definition of the field of study of geoinformatics as a system of sciences.

Geodata - structured and systematised integrated data reflecting thematic and spatial and temporal properties of objects, processes and phenomena occurring on the Earth in near-Earth space, under the Earth surface.

One of the objects of study of geoinformatics is spatial relations [50-52], which are not mentioned in computer science.

Let us make a comparison between informatics and geoinformatics in terms of applied data. In informatics data of the subject area are processed (Fig.1.3). In some cases these data are transformed into models of the subject area. For processing, models and data of any subject area are converted into computer models. For each subject area in computer science create their own data and models of the subject area. That is, interdisciplinary knowledge transfer is weakly manifested in informatics.

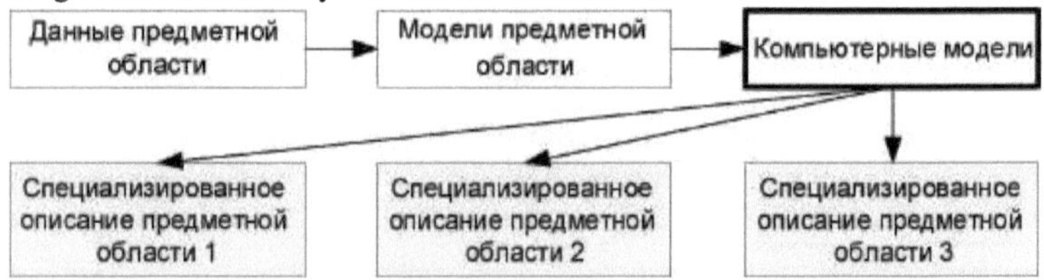

Fig.1.3 Data processed in informatics

Subject matter data are transformed into subject *matter models*. As a result, they have a specificity that reflects the characteristics of the subject area. In order that these models can be processed, they are transformed into *computer models,* which are used in computer science. For informatics, all data are equivalent, they represent the totality of the form

$$D(A1, A2, A3,.... An) \quad (1.1)$$

In expression (1.1) D - data, A - qualitatively homogeneous groups of data. As a result of processing, *specialised data sets are* obtained as a description of objects, phenomena, processes for each subject area. The paradigm of informatics application is reflected by the following chain

Subject matter data subject matter models ^ computer models specialised datasets

It should also be noted that these specialised informatics are only applicable in their subject area.

In geoinformatics, geodata are more structured than data in computer science. They initially contain three groups and have the form

$$GD = F\{(C1,C2,...Cp), (Pt1, Pt2, ...Ptm), (A1, A2, ...Al) \quad (1.2)$$

In expression (1.2) Ci is a set of coordinate (spatial) parameters (i=1...n); Pti is a set of temporal parameters (i=1...m); Ai is a set of thematic characteristics (i=1...k). This data structure makes it convenient for modelling in space and time. Fig.1.4 shows the data structure in geoinformatics.

Geodata

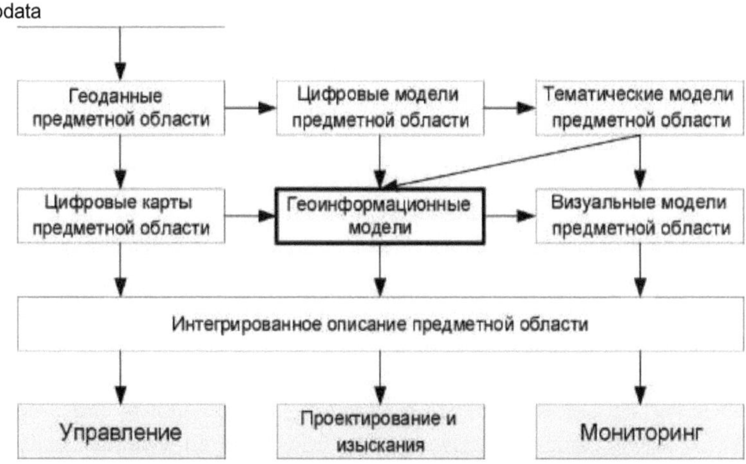

Fig.1.4 Data in geoinformatics and its transformation

It is characterised by integration. In the first stage, geodata is collected and then adapted to the subject area of the problem solving. Geoinformatics has its own research area, but like computer science, it is applied to other subject areas. This leads to the concept of subject *area geodata.*

Subject matter geodata is converted into *digital maps, digital models and subject matter topic models. The geodata is* then transformed into *geographic information models* (description models) and into an *integrated information framework* (geographic information processing framework) that contains data from *different subject areas. As* a result of the integrated *processing* using the integrated information framework, specialised *complex datasets are* obtained as a description for each subject area. The paradigm of geoinformatics application is reflected by the chain of

Geodata of the subject area geoinformation models of the
subject area integrated information framework
comprehensive datasets

The difference between geoinformatics and geodata is the integrated information basis [53, 54], which includes data from different subject areas. But the processing uses those data that are necessary. It is interesting to compare the application of topology in geoinformatics and computer science. In computer science, topology is applied in the algorithmic aspect as a tool for analysing information flows (data streams). In geoinformatics, topology is applied in the algorithmic aspect as a tool for processing geoinformation and in the spatial aspect as a tool for spatial analysis, which has nothing to do with informatics.

Geodata collection in geoinformatics is carried out by a wide range of different technologies: aerial and ground photogrammetry, geodesy, GNSS, space methods, cartography, geostatistics [55, 56], statistics and so on. In geoinformatics, there is a large variety of primary data that are preprocessed. Pre-processing is carried out in the field of geoinformatics. It includes unification of various data, correction by methods of geoinformatics and integration of data into a single environment, introduction of topology and associative relations.

Geoinformatics data used in processing are integrated data intended for use in different fields. For example, integrated data in GIS allows solving problems in cartography, photogrammetry, space imagery, cadastre, land monitoring, etc.

Geoinformatics has not only differences but also similarities with informatics. Therefore, it is reasonable to compare geoinformation and information approaches. Informational [57] and geoinformational [58, 59] approaches include the use of: information units, (geo) information models, (geo) information monitoring, (geo) information modelling, information flows.

Structurally, the geoinformation approach to the analysis of processes and phenomena is shown in Fig. 1.5. Let us consider its features [58, 59]. Allocation of three groups of data "place" "time" "topic" allows the classification of different data and their subsequent integration into a single information framework. The first three stages are not applied in informatics.

Geoinformation approach

Data grouping

Data integration
Geodata generation
Building spatial models
Visual modelling
Application problem solving
Gaining knowledge and geo-cognition

Fig.1.5. Geoinformation approach to analysing objects and phenomena

Digital models and digital maps have an integrating function, which makes it possible to combine heterogeneous information resources. The integrating function makes it possible to create a hypertext structure of the data included in the information base. Graphical and cartographic visual models in geoinformatics display a vast information space. This space includes many relations between real world objects and their attributes. Data stratification with the inclusion of a hierarchy relation means the creation of a hierarchical model, visually represented as a set of layers with common themes and attributes.

The fundamental thing in such a description is the possibility of using logical and theoretical-multiple operations to obtain new knowledge about survey objects and phenomena on the Earth's surface. Construction of spatial models includes finding spatial relations and spatial connections.

The identification of spatial relationships makes it possible to find weak and strong, explicit and implicit connections between objects located at different points in space. Visual modelling [60, 61] is key in the representation, interpretation and processing of geodata. The volumes and main complexity of geodata are so large that traditional methods of analysing (simple) information become unacceptable. Let us analyse the differences between the information and geoinformation approach.

The information approach is aimed at processing any information and solving any problems, but requires adaptation in doing so. This determines its abstractness and detachment from practice.

Geoinformation approach is aimed at processing spatial information, geodata and solving problems related to the positioning and movement of objects on the Earth's surface. It is aimed at solving problems related to the occurrence, course and disappearance of various processes and phenomena on the Earth's surface. It does not require adaptation as it is oriented to a specific domain and specific data structures.

The information approach plays the role of an intermediary in processing the raw data collected by the user and solving the tasks set by the user. Geoinformation approach plays the role of an applied tool in solving user's tasks.

The information approach is more processing-oriented, without regard to applications. This determines its instrumental character and allows it to be considered as a tool of a mediator (programmer).

Geoinformation approach is focused not only on processing, but also on generalisation and analysis of information with the target output - obtaining management information or information for decision support. This determines its integration character with

13

applications and allows to consider it as a tool for user-researcher of processes and phenomena on the Earth's surface.

The information approach appeared earlier and was focused on data processing, without regard to spatial relations. Historically, the geoinformational approach appeared later, but it takes into account the shortcomings of the informational approach and, on the contrary, adds specificity, allowing to find and use spatial relations to solve a set of problems.

The coordinate environment factor is absent in the information approach. The coordinate environment factor is present and plays an integrating key role in the geoinformation approach.

The information approach aims at identifying and modelling relationships. The geoinformation approach is aimed at identifying and utilising links and relationships [62], among which spatial relationships play a leading role.

The information approach is characterised by a mediating aspect between processing methods and applications. This defines the differentiation aspect as the main one. The geoinformation approach is characterised by an integration aspect. Thus, the geoinformation approach has its own specificity and certain advantages in relation to tasks related to spatial relations and spatial distribution of information.

In general, it should be stated that geoinformatics is a system of sciences [45] in contrast to specialised informatics. It has its own research methods and its own fields of application, different from the methods and fields of informatics. There are significant differences between them and they cannot be identified.

1.2. Comparison of space and terrestrial geoinformatics

Comparison of space and terrestrial geoinformatics is analogous to the comparison of space and terrestrial astronomy. Space geoinformatics, like terrestrial geoinformatics, is primarily related to cognition and research of the surrounding world. Considering the transfer of terrestrial research methods to the field of space research, it is necessary to distinguish the following levels: the level of information collection, the level of data, the level of technology, the level of interpretation, the level of information storage, the level of information presentation, the level of concepts, the level of generalisation, the level of cognition.

At the data collection level, there is a complete difference between space and terrestrial geoinformatics. At the data level, there is no difference between space and terrestrial geoinformatics. Both geoinformatics use geodata, which is an integrated set that includes three qualitative groups: "place" (coordinate data), "time" (temporal characteristics), and "theme" (thematic data). Geodata is a systemic information resource, which makes it possible to conduct systemic analysis and systematise the results of space research on its basis.

Space geoinformatics provides, at the data level, comparability and analysis with ground-based research methods. At the technology level, space geoinformatics is compatible with ground-based technologies and creates a tool for sharing analysis and processing methods.

At the interpretation level, Space Geoinformatics has its own processing methods, but they are fully compatible with terrestrial interpretation methods. At the levels of: information storage, information representation, concepts and generalisation, there is no difference.

At the cognitive level, Space Geoinformatics promotes the integration of sciences. However, unlike terrestrial geoinformatics, it integrates space-related sciences.

The integration of sciences in space geoinformatics is qualitatively different from the integration of sciences in terrestrial geoinformatics. Terrestrial geoinformatics integrates two blocks of sciences: Earth sciences and information sciences, including information processing. Space geoinformatics integrates three blocks of sciences: space exploration sciences, Earth science methods and information sciences, including information processing based on the geoinformation approach.

Both geoinformatics are closely related to synergetics. At the level of cognition, space geoinformatics, as well as Earth geoinformatics, draw information from the information field surrounding the human being [63-65], in what they complement each other.

Terrestrial geoinformatics gains knowledge, spatial knowledge and geo-knowledge. Space geoinformatics and space astronomy gets knowledge, spatial knowledge and space knowledge.

Thus, there are more similarities between terrestrial and space geoinformatics in terms of basic levels of development and research. In particular, there are no differences between space and terrestrial geoinformatics technologies. It follows that terrestrial geoinformatics methods are applicable to space geoinformatics, and their research fields complement each other and create a more complete picture of the surrounding world.

1.3. Remote sensing and space geoinformatics

In a broad sense, remote sensing [66, 67] (RS) is the obtaining by any non-contact methods of information about objects on the Earth's surface or in its interior. These methods are used by photogrammetry, geodesy, optics, physics, and radio engineering. In a narrow sense, remote sensing of the Earth is the acquisition of information using equipment installed on board spacecraft. Remote sensing data (RSD) is the main source for maintaining up-to-date information about objects and processes on the Earth's surface. Geoinformatics has the property of integration and integrates remote sensing methods [68, 69].

Until recently photogrammetric methods of remote sensing were the most widespread. This is due to the high level of technical, methodological and technological development of the processes of photo-image acquisition, but most importantly due to their high information content. This characteristic should be distinguished from the information capacity of a photo carrier or information volume of a file, which is obtained from a photo image.

In remote sensing it is possible to obtain information about the object of study in different spectral ranges: X-ray, ultraviolet, visible, infrared. The shorter the wavelength, the higher the accuracy of measuring the position of the object. Optical wavelengths are shorter than thermal or radar wavelengths.

Therefore, optical observations recorded on photographic film or with scanning devices are more informative and accurate. The complexity and peculiarity of remote sensing is determined by the significant influence of interference on the useful signal. The observer in remote sensing can fulfil a passive or active role.

The diversity of information in remote sensing requires its systematisation. The basis for such systematisation in classical geoinformatics is geodata. However, in space geoinformatics the term geodata should be replaced by the term "integrated space data" (ISD). ICDs have similarities and differences with both geodata (GD) and SDD. Their similarity to geodata lies in the principle of organisation, i.e. the acquisition of an integrated, systematic set of data as a complex system. The difference is that the basis for the integration of ICDs is not data from the Earth's surface, but spatial data from outer space.

With SDI, the similarity of ICDs lies in the technologies used to obtain them, while the difference lies in the way they are organised. As information collections, DLDs are collections of different data produced by different technologies and used in diversified technologies [70]. These diversified DDD data are used in specialised analysis and processing technologies. In contrast to diversified SDD, integrated SDIs are used in space geoinformatics. This integrated data is a single data system that includes all types of RS data integrated into a single information model.

Diversified SDRs are homogeneous for one technology, but more different when compared to each other for different technologies. This leads to the need to change data formats and even processing methods when transferring RSD from one diversified technology to another. ICDs are more heterogeneous (heterogeneous), which makes it possible to consider them as data federations rather than data collections. However, for different technologies ICDs practically do not change. This significantly simplifies the use of such data in different processing technologies and different processing complexes. The integration of ICDs increases the possibility of information exchange and creates conditions for interdisciplinary transfer of models and knowledge.

1.4.   Geodetic astronomy and space geoinformatics

For the period of more than a hundred years [71] Geodetic astronomy has passed a long way of development [72-74] and occupies a priority place among the complex of major scientific disciplines. Astronomy is one of the oldest sciences. Ancient civilisations left behind numerous astronomical artefacts confirming their knowledge of the regularities of motion of celestial bodies. Historically astronomy included astrometry, navigation (by stars), observational astronomy. Navigation is of great importance for solving problems of transport control and movement. Nowadays, professional astronomy includes radio astronomy [75]. Astronomical determinations together with the results of geodetic and gravimetric measurements allow: to establish initial geodetic dates; to provide orientation of the reference ellipsoid axes in the Earth's body; to determine the parameters of the Earth's ellipsoid; to determine heights relative to the reference ellipsoid. The need for geodetic measurements, comparison of astronomical measurements with terrestrial ones, transformation of astronomical measurements into terrestrial ones - led to the emergence of geodetic astronomy.

Features of geodesic astronomy. If we consider geodesic astronomy as an academic discipline, then as a result of studying this discipline students should know:

the geometry of the celestial sphere, the mechanics of the daily motion of the stars; coordinate systems and time measurement systems [76];

The main problems and solutions of astrometry;

organisation of Services for determination of Earth rotation parameters and pole coordinates;

the theory of astronomical coordinate reductions;

the creation of star catalogues;

The theory and practice of astronomical definitions;

precise methods of determining astronomical coordinates and azimuths, their purpose

When choosing a reference frame, the sphere of observation is selected [77, 78]. Depending on the choice of the centre of the celestial sphere, there are: topocentric celestial sphere - the centre is on the surface of the Earth; geocentric celestial sphere - the centre coincides with the centre of mass of the Earth; heliocentric celestial sphere - the centre is combined with the centre of the Sun; barycentric celestial sphere - the centre is in the centre of gravity of the Solar System.

To define a spherical coordinate system on the sphere, two mutually perpendicular great circles are chosen, one of which is called the main circle and the other is called the initial circle of the system. The following spherical coordinate systems are used in geodetic astronomy: horizontal coordinate system; first and second equatorial coordinate systems; ecliptic coordinate system. The name of the systems usually corresponds to the name of the great circles taken as the basic circle.

In geodetic astronomy, the astronomical latitude and longitude, f and X, and the astronomical azimuth of direction A are determined.

Astronomical latitude f is the angle between the plane of the equator and the plumb line at a given point. Latitude is counted from the equator to the north pole from 00 to +900 and to the south pole from 00 to -900.

Astronomical longitude X - is defined as the dihedral angle between the planes of the initial and current astronomical meridians. Longitude is counted from the Greenwich meridian to the east (Xe- east longitude) and to the west (Xw- west longitude) from 00 to 1800 or, in hour measure, from 0 to 12 hours (12^h). Sometimes longitude is reckoned one way from 0 to 3600 or, in hourly measure, from 0 to 24 hours. Astronomical azimuth of direction A - is defined as the dihedral angle between the plane of the astronomical meridian and the plane passing through the plumb line and the point to which the direction is measured.

Geodetic astronomy uses systems of sidereal and solar time based on the rotation of the Earth around its axis. This periodic motion is highly uniform, not limited in time, and continuous throughout the existence of mankind. In addition, astrometry and celestial mechanics use the systems of ephemeral and dynamical time as an ideal construction of a uniform time scale; the system of atomic time is a practical realisation of a perfectly uniform time scale. The peculiarities of using the results of astronomical determinations are as follows.

Determination of the components of the plumb line slope from astronomical observations allows to establish relations between geodetic and astronomical coordinate systems;

Astronomical determinations of azimuths of directions to terrestrial objects control angular measurements and limit random and systematic errors in angular measurements;

Astronomical determinations of geographic coordinates provide a means of determining the positions of objects moving relative to the earth's surface at sea and in the air. It is an important factor in the management of transport systems;

Astronomical determinations of geographical coordinates and azimuths of directions are used to control angular measurements in polygonometric runs and other angular constructions, in topographic and geodesic support of troops.

Methods of astronomical determinations are divided into exact and approximate. Precise methods are those that allow, with the current state of the theory of geodetic astronomy and its instrumental base, to obtain with maximum accuracy the values of latitude, longitude and azimuth of directions. Approximate methods determine astronomical coordinates with an accuracy of 1" to 1'. The general distinguishing features of approximate methods are: direct measurement of observed values, a small number of observation techniques, fixing the moments of observation no more accurately than 1s, frequent use of the Sun as an object of observation, the use of simplified observation techniques and simplified processing formulae.

Space geoinformatics as an integrator of sciences. Space geoinformatics as a new scientific direction has emerged and is developing on the basis of a set of integrations of various scientific directions. As the first type of integration we should note the integration of geoinformatics with remote sensing technologies. As the second integration we should note the integration of transformed Earth sciences into space disciplines (left column of Fig. 1.6) into a single complex. The basis of integration in this direction is geodetic astronomy as the most historically developed science among the considered sciences. As the third integration we should single out the integration of space disciplines (space photogrammetry, radar images processing, thermal images processing, lidar methods) into a single complex. All these three complexes, in turn, have been integrated into space geoinformatics. Space geoinformatics contributes to

It is characterised by an integrated approach to the study of the world around us. Figure 1.6 shows the integration of various sciences in space geoinformatics.

The right column shows the Earth sciences, the left column shows their transformation into space exploration.

Figure 1.6: Integration of sciences in space geoinformatics.

Space geoinformatics, like geodetic astronomy, provides comparability and analysis at the data level. But geodetic astronomy solves communication problems between astronomy and geodesy. Space geoinformatics solves communication problems for more sciences.

Space geodesy serves as the basis for the transfer of geodetic measurement methods to space. In Figure 1.6, the left column shows the integration of sciences into space geoinformatics. The right column denotes the sciences that have transformed into space sciences to perform research within space geoinformatics. Comparative planetology [7882], which differentiates the research direction of small celestial bodies, stands out.

At the technology level, Space Geoinformatics integrates technologies and methods for analysis and processing. At the same time, it is a tool for interdisciplinary knowledge transfer. At the level of cognition, Space Geoinformatics, similar to terrestrial geoinformatics, promotes the integration of sciences.

As a cognitive tool, space geoinformatics extracts information from the information field, studies and creates spatial knowledge, including geo-cognition. As a means of forming a picture of the world, space geoinformatics complements other sciences and scientific directions.

The information component of modern society is the basis of development. The importance of space geoinformatics and geodetic astronomy lies not only in information processing, but in the extent to which the model of the world and society is expanded. The significance of space geoinformatics in the sphere of scientific research lies in the extent to which new models of space geoinformatics are adequate to the real human environment and contribute to the development of civilisation.

Considering the process of space exploration as a process of cognition of the world, we can consider that Cosmic Geoinformatics extends the space of Earth geoinformatics research to the space of geodesic astronomy. If we consider the system of nested spaces (Fig.1.7) given in [77], then Space Geoinformatics covers all spaces.

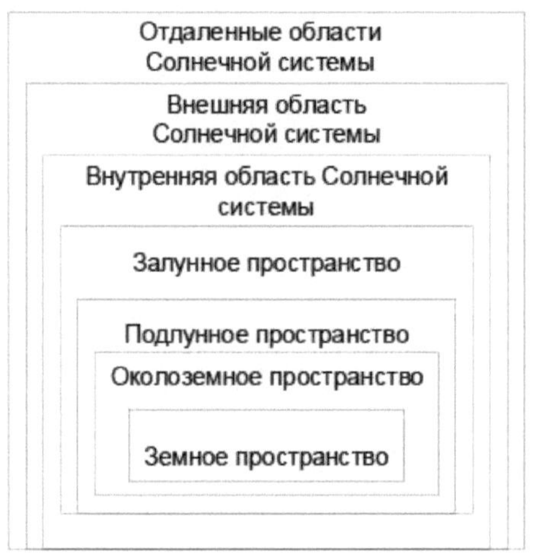

Remote areas
solar system
External area
solar system
The inner region of the solar
system
Zalun space
Sublunar space
Near-Earth space
Earth space
Fig.1.7. Space as a set of nested spaces.

Space astronomy covers spaces other than terrestrial and near-Earth spaces. The research methods of geodetic astronomy and space geoinformatics have transcended terrestrial boundaries and extended to the exploration of outer space.

1.5. Comparative planetology and space astronomy

Comparative planetology is one of the components of space astronomy and space geoinformatics. The object of study of comparative planetology [84-90] are planets mainly of the solar system. This means the application of comparative analysis methods in the research. In comparative planetology, the main object of comparison or the first system of comparison is the Earth. It is the most studied planet on which various studies and hypothesis testing can be carried out. The use of Earth studies is most common in sciences such as planetary geology, geomorphology and atmospheric sciences. Comparative planetology transfers Earth science methods (geoinformatics, geodesy, geodynamics, photogrammetry, cartography) to other planets and objects in the solar system. Many Earth sciences provide the basis for the study of other planets and space objects. Space geoinformatics also transfers the methods of terrestrial geoinformatics and Earth sciences to the field of space research, not only of planets, but also of other celestial bodies

For a long time, the content of comparative planetology was based on the geological component. In space astronomy, the content was based on astronomy, Earth science methods and computer science methods.

In comparative planetology, empirical methods for the study of planets are widely presented [81]. Comparative planetology studies the state and development of small celestial bodies [82]. Space geoinformatics also studies planets, but in the aspect of their spatial models and real motion trajectories.

The search for life in outer space [83] and "Earth-like" planets [84] is becoming a research area of comparative planetology. Due to the emergence of the scientific fields of space geodesy and space geoinformatics, the number of components of comparative planetology has increased.

The second system of comparison in comparative planetology is the Solar System. The space of the solar system can be considered as a complex system from a systemic point of view. It serves as a "supersystem" for the Earth and near-Earth space. Hence the need to apply the methods of system analysis in comparative planetology and in the study of outer space.

The complexity of in-situ experiments motivates the wide application in comparative planetology of various types of modelling: formal, analogue, cartographic, computer, stochastic. Space geoinformatics also applies different types of modelling. In particular, virtual modelling has recently been widely used in many sciences. It allows carrying out non-contact experiments

The peculiarity of comparative planetology is that it has essentially transformed into a scientific complex, rather than being reduced to one narrow science. Comparative planetology is defined as a complex of sciences studying: planets, planetary satellites, the Solar System and other planetary systems. Its scope of study includes objects ranging from micrometeorites to gas giants. Planetology research method studies physical properties, chemical composition, structure of the surface, inner and outer shells of planets and their satellites, as well as the conditions of their formation and development. The peculiarity of space geoinformatics is also that it has transformed into a scientific complex, rather than being reduced to a single science.

Planetology as a scientific discipline integrates many disciplines: planetary geology (together with geochemistry and geophysics), physical geography (geomorphology and cartography, as applied to planets), atmospheric sciences, theoretical planetology and exoplanet studies. A number of disciplines are related to comparative planetology, such as space physics, astrobiology and sciences that study the influence of the Sun on the planets of the solar system.

Space geoinformatics as a scientific discipline integrates many sciences. Global space monitoring is developing as an application of space geoinformatics. Information support of data processing in space geoinformatics and comparative planetology is developing. The problem of big data, voiced in 2005, has existed in comparative planetology for a very long time. It should be noted here that the problem of big data includes not only large amounts of information, but also information that is poorly structured and not systematised. It is this type of information that is found in comparative planetology. Due to the emergence of the big data problem, the methods of information analysis have significantly expanded, which favourably affects the development of comparative planetology. This makes comparative planetology an integrative science and requires the study of a number of supporting disciplines and areas for its mastery.

One of the support tools and sources of information for comparative planetology is global monitoring. The concept of "globalisation" is used by various authors to refer to a wide range of phenomena and trends. Global monitoring is the monitoring of global processes occurring both on the Earth's surface and in near-Earth space, as well as beyond near-Earth space. The basis of global monitoring is space monitoring integrated with geoinformation monitoring.

In comparative planetology there are quite a large number of terms and sciences related to the Earth, although we are talking about other planets. For example, geoinformatics has a development in the form of space informatics. There is a geography of extraterrestrial territories. Geography: (dr. Greek ugshurafSa, earth writing, from uz - Earth and uoshra[1] - writing, describing). Geography of extraterrestrial territories sounds like "land writing of extraterrestrial territories". It is quite appropriate to use the term "planetography" and "cosmography" instead of "geography of extraterrestrial territories", which is already done by individual scientists. The same applies to space geodesy. There is no Earth in outer space. But a suitable term has not yet been found. The term for the science that measures other planets is suggested to introduce the term "planetometry" by analogy with geometry.

1.5.1 Systematics of planetary orbits

In comparative planetology and space geoinformatics, the expression spatial comparison is rarely used because it is often qualitatively characterised. For precise metric comparisons, the term geometric comparison is used, meaning quantitative comparison or proportional comparison. Comparative planetology provides a good reason to compare patterns in different planetary configurations. The solar system is a planetary system that includes the Sun and all natural space objects orbiting around it. The solar system is part of the Milky Way galaxy. The system includes eight planets and objects called small bodies of the solar system. From the perspective of system analysis, the Solar System is a model of a complex system with systemic features.

Any spatial system is created by the prince of homogeneity and stability of the properties of the system elements or regularity in their arrangement. Using the comparative method, let's investigate the systematisation of orbits of planets of the Solar system. When systematising the orbits of the Solar System planets, the astronomical unit [85] was taken as a basis, and the systematics was performed in linear measures of distance from the Sun. The results of the study are presented in Table 1.1. It gives conditional (comparative) presented radii (R) of orbits of different planets and their number or order in terms of distance from the Sun.

Table 1.1. Radii of planetary orbits expressed in relative measures

N	R	Mercury Y.	Venus	Earth	mars	Jupiter	Saturn	uranium	Neptune
1	Mercury Y.		1	0,5	0,2	0,25	0,07	0,0	0,020	0,013
2	Venus	1,9		1	0,6	0,47	0,14	0,1	0,037	0,024
3	Earth	2,6	1,4	1	0,66	0,19	0,1	0,052	0,033
4	Mars	4,0	2,1	1,9	1	0,29	0,2	0,079	0,051
5	Jupiter	13,7	7,2	11,9	3,42	1	0,5	0,271	0,173
6	Saturn	25,1	13,3	9,53	6,28	1,83	1	0,496	0,317
7	Uranus	50,6	26,7	19,14	12,64	3,70	2,0	1	0,639
8	Neptune	79,1	41,8	30,03	19,78	5,78	3,2	1,564	1
9	KCor.	0,886	0,886	0,886	0,886	0,886	0,886	0,886	0,886
10	KRegr.	10,32	5,4	3,927	2,58	0,75	0,41	0,20	0,13

For example, in the Mercury column, a unit corresponds to the radius of Mercury's orbit and all other planetary orbits are measured in that unit. In the Neptune column, the unit corresponds to the radius of Neptune's orbit and all other planetary orbits are measured in this unit. Naturally, the largest values in the Mercury column and the smallest values in the Neptune column.

Line 9 calculates the correlation coefficients (KCor) between orbital radii (R) and planet number (N) from 1 to 8. The presence of a large correlation coefficient indicates the presence of a relationship or pattern. The same correlation coefficient for different measures indicates the stability of a single pattern of arrangement and its independence from the chosen linear measure of orbits. For all planets the correlation coefficient (N/R) (line 9) is the same. This indicates that there is a common systematics in the arrangement of all planets.

Line 10 calculates regression coefficients, i.e. the linear slope of the dependence $R=f(N)$ $R=K\,N+h$. The change of K depending on the unit of measurement indicates that the dependence rather has a non-linear character. Fig.1.8 shows the dependence R=fN)

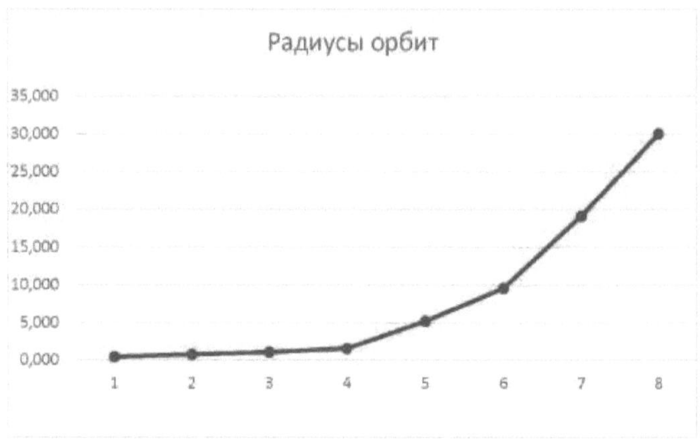

Fig.1.8. Dependence of change of radius of orbit on number of a planet

In Fig. 1.8, the vertical line shows the value of the planet radius in units equal to the Earth's orbital radius. It can be seen that for the first four planets the dependence is close to linear, for the following planets the dependence is nonlinear.

If we assume that the radii of the first 4 planets depend linearly on their ordinal number, we can calculate the correlation for these planets. For the first 4 planets, the correlation coefficient turns out to be 0.991 and the regression coefficient 0.368. This indicates that for the first 4 planets a linear relationship is acceptable and the regression coefficient of this line (slope) is 0.368.

For the remaining 4 planets (5-8) it is necessary to consider a nonlinear dependence. We have chosen variants for $R=f(N)$: quadratic dependence $R=AN^2$ and fractional degree dependence $R=AN^{3/2}$. Here A is some constant/ For the quadratic dependence of planets from 5 to 8 (Table 1.1), a correlation coefficient of 0.94 was obtained. For the fractional degree dependence $R=AN^{3/2}$, a correlation coefficient of 0.963 was obtained, i.e. higher. Both coefficients are higher than in Table 1.1 - 0.886.

Different correlation coefficients speak, as it were, about different mechanisms of orbital formation for planets with numbers 1-4 and planets with numbers 5-8. But the mechanism should be unified. In this case it is reasonable to use the exponential dependence for all planets. the exponent behaves at small values of the argument as a straight line. But at higher values it shows nonlinearity. The exponential at small values of the argument is close to a straight line, and at large values it shows nonlinearity. The model $R=Ae^N$ was chosen. The correlation coefficient for all 8 planets turned out to be equal to 0. 962, which is higher than 0.886. Conclusion The orbital radii of the solar system planets obey the exponential model.

1.5.2 The Earth-Moon system.

The peculiarity of the study of space objects is their interrelation. Objects influence each other and heavy objects influence small objects and the trajectories of passing space bodies. There are no independent bodies in space. In fact, there are systems of connected bodies in space. One such system is the Earth-Moon system, which significantly affects the life activities of mankind [86]. Analyses of most systems begin with their structure. Analyses of spatial relations and spatial interactions. In modern conditions of computer modelling of spatial processes, spatial interactions are modelled by information interactions. In addition, a convenient form of describing spatial processes is the information situation model. In spatial studies of dynamics, the information situation fulfils the functions of a photograph fixing the state of objects for a certain period of time. This allows analysing further behaviour of objects or spatial systems of the Earth-Moon type

Dynamics of the Earth-Moon system. Many popular science reference books and even some textbooks say that the Moon revolves around the Earth. This is an inaccuracy. From the course of physics we know that any connected system of two or more bodies has a common centre of mass. In the Earth-Moon system (EMS) the common centre of mass of the EMS (barycentre) [87] is located about Rb=4700 km from the centre of mass of the Earth. The Moon rotates around this centre of mass. A more precise expression is: the Moon revolves around a common centre of mass or barycentre with the Earth, which is 4700km away from the Earth's centre of mass.

Each revolution takes 27.3 Earth days and is called a sideric month. On average, the Moon is about 60 Earth radii away from the centre of the Earth, which is 384405 km. Figure 1.9 shows a diagram of the NWL.

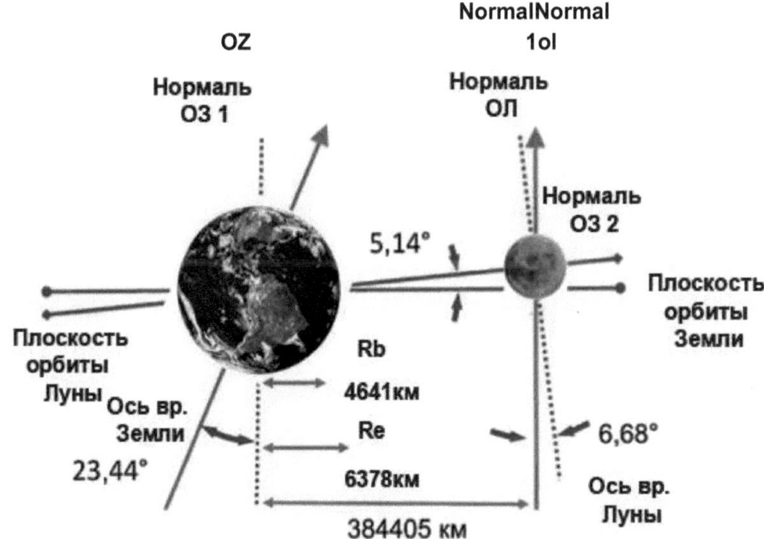

Fig.1.9 Earth-Moon system in orbital aspect

Figure 1.9 shows that the Moon's orbit is inclined by 5.14° with respect to the plane of the Earth's orbit. Two normals can be drawn to the plane of the Earth's orbit. One normal (normal 031) passes through the centre of mass of the Earth the second normal (normal

032 Fig.1.9) passes through the centre of mass of the Moon. The distance between the normals characterises the distance between the centres of the Earth and the Moon. The Earth's rotation axis is deviated from normal 031 by 23.44°. The axis of rotation of the Moon is deviated from the normal 032 by - 6.68°. The conditional point of intersection of the axes of rotation of the Moon and the Earth has coordinates 35446 km from the centre of mass of the Earth and 81754 km normal to the plane of the Earth's orbit. The Moon rotates around the barycentre on an ellipse with the parameters given in Fig.1.10. The average eccentricity of the moon's orbit is 0.0549006. This gives reason to consider it close to circular.

Fig.1.10. Parameters of the Moon's orbit.

The motion in the Earth-Moon system is shown in Fig. 1.11. The common centre of mass of the NWL or barycentre is designated BC. The centre of mass of the Moon is designated CML. The centre of mass of the Earth is denoted by CMZ. Figure 1.11 shows the directions of rotation of the Earth and the Moon around their axes. The dotted line indicates the rotation
of the Earth's centre of mass around the barycentre.

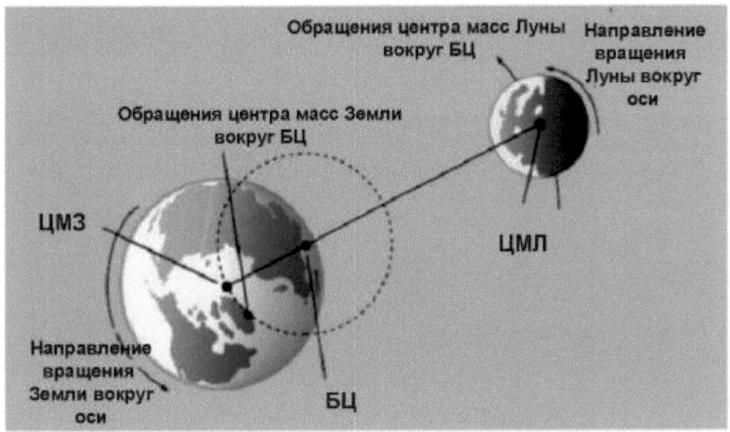

Fig.1.11. Motion in the Earth-Moon system.

Another peculiarity should be noted. Some reference books claim that the tides are influenced by the Moon. This is inaccurate. The Moon influences, but by means of the barycentre. Tides occur as the Earth rotates on its axis, but through the influence of the barycentre.

Libration points in the Earth-Moon system

There are phenomena related to the motion of the Earth and the Moon in the NWL. These include Lagrange points [88], which are also called libration points [89, 90]. Let us consider a dynamic information situation for which the spacecraft (SC) is on a line connecting the Earth and the Moon (Fig.1.12). When travelling along this line, the Earth's gravity, the Moon's gravity, and the centripetal acceleration due to the rotation of the line act on the spacecraft. It can be hypothesised that there is a point at which these three accelerations are nullified. It turned out that such points exist in NWL and there are five such points (Fig.1.12). they were called libration points and denoted by the symbol L.

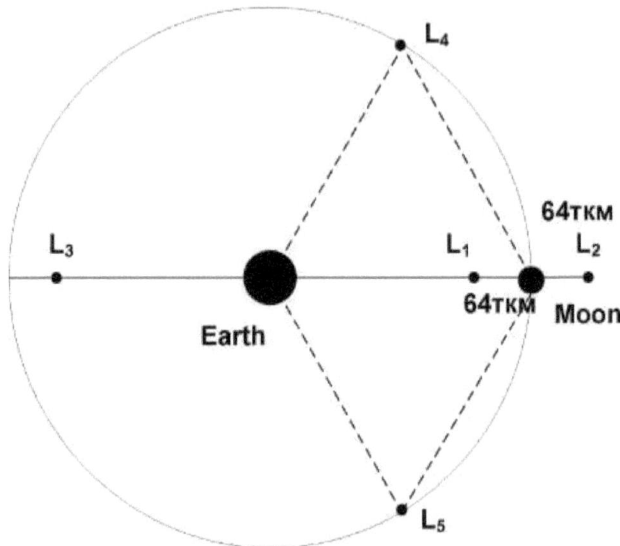

Fig.1.12. NWL libration points

The libration points represent a partial solution of the three-body problem. In this problem, the orbits of all bodies are circular, and the mass of the third body is much smaller than the masses of the other two bodies. In this case, two massive bodies are assumed to orbit around their common centre of mass (barycentre) with constant angular velocity. The solution shows that there are five points in the space around the two bodies at which the third body with small mass can remain stationary in the rotating frame of reference (ROR). At the libration points, the gravitational forces acting on the small body are balanced by the centrifugal force.

Figure 1.12 shows five libration points. It is fundamental that this system is mobile and Fig. 1.12 captures, as a photograph, one of the possible states of the NWL. Three of the libration points are on the line connecting the Earth and the Moon (LZL) and are called collinear libration points. The first is between the Earth (Earth) and the Moon (Moon) and is labelled Li, the second libration point is behind the Moon - L2, and the third collinear libration point - L3 is on the back side of the Earth with respect to the Moon. The fourth L4 and fifth L5 libration points are on two sides outside the LZL. They are called triangular libration points. The L4 and L5 points are called triangular or Trojan points. Points Li, L2, L3 are points of unstable equilibrium, at points L4 and L3 equilibrium is stable

The conditionality of the model in Fig. 1.12. is that the orbit is not strictly circular. Therefore, the study and search for libration points is a scientific problem. If we place a small body in any libration point, it will remain there within the framework of such a simple system. From the physical point of view, the libration point is like a potential hole.

In a coordinate system with the origin at the centre of mass of the system and with the axis directed from the centre of mass to the less massive body, the coordinates of these points in the first approximation by "*a*" *are* calculated using the following formulas:

$$r_1 = (R\,[1-(a/3)^{1/3}],0)$$

$$r_2 = (R\,[1+(a/3)^{1/3}],0)$$

$$r_3 = (-R\,[1+(5a/12)],0)$$

$a=M_2/(M_1+M_2);\ \ R$ is the distance between the bodies; M1 is the mass of the more massive body; M2 is the mass of the second body. If M2 is much less in mass than M_1, then the points L_1 and L_2 are at approximately the same distance r from the body M2, equal to the radius of the Hill sphere

$$r \approx R(M_2/3\,M_3)^{1/3}$$

where *R is the* distance between the components of the system

If a small body or spacecraft leaves these points, it can move along periodic orbits (halo-orbits, Fig. 1.13) in their vicinity [91]. In Fig. 1.13, the distance between the Moon and the Earth is chosen as a conventional unit. The line Earth (Earth) - Moon (Moon) - (LZL) is selected. The coordinate (dynamic) system is chosen so that one axis (horizontal) is located along the LZL direction, the other axis (vertical) is normal to the LZL.

A small body or spacecraft can move around the libration point along such orbits. For libration points L1, L_2 of the Earth-Moon system, the period of motion along these orbits will be of the order of 12 - 14 days.

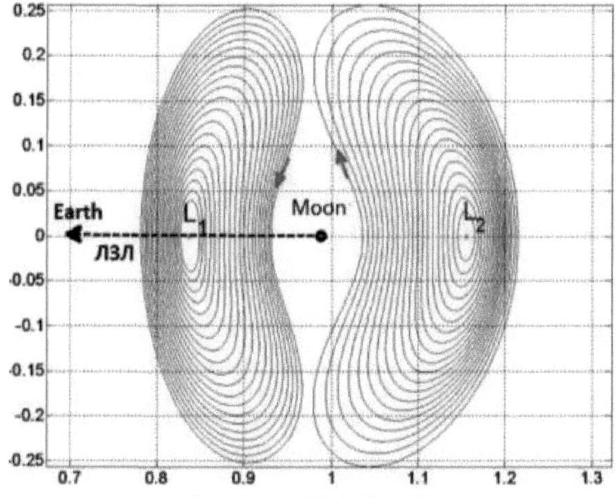

Fig.1.13. NWL halo-orbits

Estimates show that the sizes of Halo orbits are about 0.2 along the LHL and about 0.5 normal to the LZL. An important fact should be noted. At first sight homogeneous space in the SPL becomes inhomogeneous when gravity and rotation of large bodies are taken into account. Therefore, spacecraft trajectories should be chosen so that they do not fall in the region of libration points. The libration points exist not only in the NWL, but also in the Sun-Earth system. Therefore, spacecraft motion trajectories should be chosen taking into account these points. The libration points exist in the Sun - Earth group planet (Mars) system. Therefore, halo-orbits can be used to observe satellites of planets as relatively stable trajectories. Relativity is due to the fact that the spacecraft will be affected by the Sun's gravity, the pressure of sunlight.

For points L4, L5 there are the following estimates

$$r_4 = (\frac{R}{2}b, \frac{\sqrt{3}R}{2})$$

$$r_5 = (\frac{R}{2}b, -\frac{\sqrt{3}R}{2})$$

The libration point L_1 is convenient for the location of near-lunar space stations there. If the station is located in the libration point, the transition from the libration point to the halo-orbit allows flying to any point on the Moon surface. It should be noted that the presence of libration points is a characteristic of the presence of energy tunnels in space. These tunnels ensure minimisation of energy expenditure during space travel.

The Sun-Earth system influences the Earth-Moon system. Table 1.2 summarises the characteristics of the two dual systems.

Table 1.2: Characteristics of dual systems.

system	Sun-Earth	Earth-Moon
Mass of the larger body M (kg)	Sun 1.991×10^{30}	Earth 5.98×10^{24}
mass of the smaller body m (kg)	Earth 5.98×10^{24}	Moon 7.35×10^{22}
Mass ratio m / (M + m)	0,000 003	0,012
Distance M - m	149,600,000 km 1 AU	384.401 kilometres
Distance M - SM	440 kilometres	4667 km
Distance m - SM	149.599.551 km 0.999 997 AU	379734 km
Distance Li - m	1,501,557 km 0.010 037 AU	64499 km
Distance L2- m	1,491,557 km 0.009 970 AU	58.006 kilometres
Distance L3- M	149,599,737 kilometres 0.999 998 AU.	381.678 kilometres

Table 1.2 shows the characteristics of the two dual *Sun-Earth* (SWE) and *Earth-Moon* (EEM) systems. It draws attention that the GCC libration points L1, L2 are located in the body of the Earth. This may lead to the specificity of halo orbits.

1.6 Space geodesy and space geoinformatics

To compare space geodesy and space geoinformatics, it is reasonable to compare conventional geodesy with space geodesy and conventional informatics with space geoinformatics. Figure 1.14 shows a typical scheme of ground geodesy. The main purpose of terrestrial geodesy is to investigate the terrestrial information field and makes measurements of terrestrial objects.

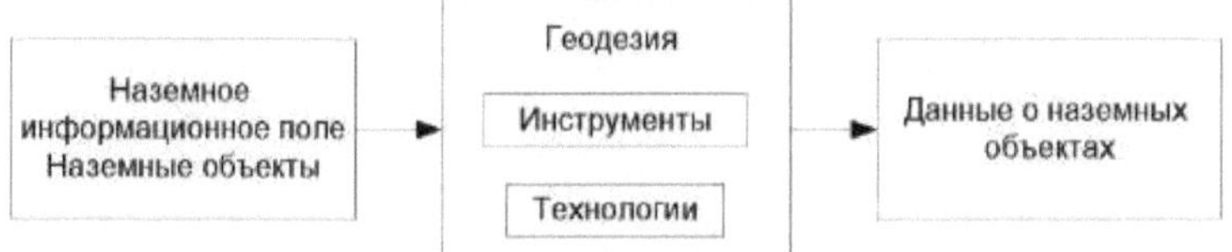

Fig.1.14. Geodesy as a tool for researching the terrestrial information field

Ground geodesy or simply geodesy uses geodetic instruments and tools, and geodetic technologies. On the basis of measurement and processing of measurements from a set of parameters of the information field select (allocate) necessary for description of objects of research and processes in which they participate.

Figure 1.15 shows a typical scheme of space geodesy. The main purpose of space geodesy is to investigate the space information field and conducts measurements of extraterrestrial objects.

Fig.1.15. Space geodesy as a tool for space information field research

The term "extraterrestrial objects" includes natural space objects and human-made objects launched into outer space. Natural space objects and passive man-made objects travel along trajectories that can be determined by the laws of mechanics under the known effects of space fields. The term "extraterrestrial objects" also includes artificial space objects that move in space according to the laws of Earth's three-dimensional mechanics and in violation of those laws. For example, they can move with a trajectory with violation of smoothness or interrupt their motion in one point of space three-dimensional space and appear in another. Comparing the schemes in Fig. 1.14 and Fig. 1.15 we can state that space geodesy differs from non-terrestrial geodesy in the first place in the vows of research and in the second place in the technology of calculations, that is, the use of space coordinate systems.

When moving from geodesy to geoinformatics, information collection subsystems should be distinguished. Geodesy is one of the collection technologies that are applied in geoinformatics. Figure 1.16 gives a simplified model of collection technologies.

Fig. 1.16. Information acquisition technologies used in geoinformatics and space astronomy.

Figure 1.16 shows that geoinformatics uses more different collection technologies. Geodetic collection technologies are part of the collection technologies in geoinformatics. Figure 1.17 gives a diagram of terrestrial geoinformatics.

Fig.1.17. Transformation of source information into information result in terrestrial geoinformatics.

Unlike geodesy, which results in data, geoinformatics results in models, solutions to applied problems and new knowledge. Figure 1.18 shows the scheme of space geoinformatics.

Fig.1.18. Transformation of source information in space geoinformatics and astronomy.

Space geoinformatics, like space astronomy, differs from its terrestrial analogue first of all by changing the object of research. The second difference of space geoinformatics consists in the application of special processing and analysis methods, conditioned by specific conditions of research objects. Space geoinformatics, like terrestrial geoinformatics, uses information-measuring systems [92].

1.7 Space geoinformatics and space astronomy
as a scientific complex

Space astronomy can be considered as a complex system. As such, geoinformatics can also be considered as a systemic scientific complex. Space geoinformatics emerged from the integration of geoinformatics methods with remote sensing technologies and space research methods. Any new science does not contradict scientific development. It is consistent with the existing system of sciences and has a certain place in this system. Fig.1.19 shows the relation of space geoinformatics with close scientific directions.

Space geoinformatics includes remote sensing and terrestrial geoinformatics in terms of concepts and methods.

Applied geoinformatics, as an application for solving terrestrial problems, is outside the domain of space geoinformatics. A significant part of geodesy and photogrammetry is also included in space geoinformatics at the level of theory and technology. The terrestrial applications of these sciences are outside the realm of space geoinformatics. Geodesy, in interaction with space research, is transformed into space geodesy. Space geoinformatics overlaps significantly with the field of space research. In the field of space geoinformatics falls the study of the Earth from space and the study of small celestial bodies.

Space\ geoinformatics

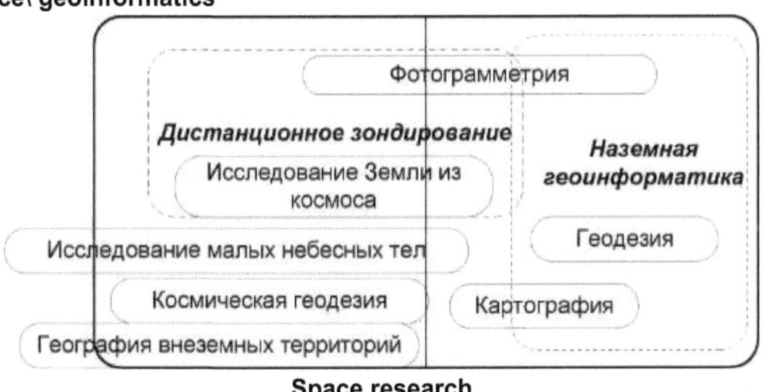

Space research

Fig.1.19. Space geoinformatics as a scientific complex

Terrestrial geoinformatics integrates cartography and geodesy techniques for three-dimensional modelling and cartographic description of spatial objects. This creates a convenient tool for the study of small celestial bodies, which includes photographing them, building three-dimensional models and constructing cartographic works of these bodies. This also creates the conditions for research in the field of extraterrestrial territories by space geoinformatics methods. Earth exploration from space is also methodologically covered by space geoinformatics.

Not all scientific directions are shown in Fig.1.19. Therefore, it is necessary to mention them. The area of space geoinformatics partially includes the scientific direction of asteroid and comet hazard research. Space geoinformatics includes the direction of comparative planetology. It fully includes the direction of space monitoring. In this direction should be divided: monitoring of outer space and global monitoring of the Earth from space. As a science investigating real space and linking different spaces, space geoinformatics uses and develops different coordinate systems. As a cognitive tool, space astronomy extracts information from the information field.

2. Modelling in space astronomy

Modelling in space astronomy is based on information modelling [93-96] and geoinformation modelling [97-99]. Modelling [100] is a method of scientific cognition in which the object under study is replaced by a model. This model is influenced and its behaviour under the influence of the influence is studied. A model can represent an entity: an object, its part (e.g., structure), its state. This is an object model. A model can represent a process. This is a process model. A model can represent a property. This is an attribute model. A model can structurally correspond to a system. It is a system model. Modelling depicts an action. Therefore, models and modelling complement each other as tools for studying and learning the picture of the world [101]. Modelling as a cognitive technique is used to obtain knowledge [102]. In space astronomy, special digital modelling [103] and special information systems [104] are used.

Modelling is a form of reflection of reality [105, 106] was used in the ancient era. Implicitly, modelling was used in the Renaissance. It is known that many Italian architects used models for the structures they designed. In theoretical works of G. Galileo and Leonardo da Vinci not only models are used, but also the limits of applicability of modelling are clarified. I. Newton uses the method of modelling quite consciously Since the XIX century modelling is clearly included in many areas of science. The basis of modelling is the construction of a model with fixed and free variables. Free variables are divided into independent and dependent variables. Modelling itself is the process of obtaining a result, in which a change in the independent parameters of the model entails a change in the dependent parameters of the model. At the same time, a change in the model parameters can lead to a change in its state. Thus, modelling leads to changes in the parameters and state of the model. Further, the change of parameters and state of the model can be transferred to the change of parameters and state of the modelling object. The set of model states allows drawing conclusions about the behaviour of the object under study.

In addition, modelling is the process of obtaining a result under certain conditions or in a certain situation. Changing the conditions or situation changes the result of modelling. One of the purposes of modelling is to study and predict the behaviour of an object under given conditions. It allows to investigate the stability of the object under external influences. Modelling allows to recreate the processes of behaviour of the object of modelling with less expenses and, if necessary, to reveal the criteria of optimisation of such behaviour. Naturally, it is difficult to recreate a model that meets all the characteristics of the object, so the modelling uses the essential characteristics of the object, which introduces a certain subjectivity. It is not always acceptable to create material models, so it is more effective to use abstract modelling.

Three types of modelling processes are considered to be the main varieties of modelling processes: mathematical, in-situ and information modelling. In full-scale modelling, the primary object under study is replaced by a similar secondary object-analogue corresponding to it by its main characteristics. This simplified object, or analogue model, reproduces the properties of the primary object under study while preserving its physical nature. The possibilities of analogue modelling are limited. In many areas of research, full-scale experimentation is impossible because it is either prohibited due to unpredictability, or dangerous, or not feasible on a large scale.

In information modelling, the primary object under study is replaced by an information model. In mathematical modelling, the primary object under study is replaced by a formula. Mathematical modelling is based on the use of a mathematical model describing the state of the modelling object. It requires a strictly analytical description, an accurate mathematical model. adequate to the real process and object. In the presence of uncertainty or incompleteness such a model is not applicable.

Information modelling is based on "mirroring" of real objects in the information field. Information modelling can be considered as coding of a real object and its description in the information field. From these positions an information model [107] can be considered as a coded message. Therefore, in many cases it is preferable to use information modelling. Modelling can have different forms and purposes. For example, the simplest modelling is related to partitioning into two classes in parameter space using a separating hyperplane [108]. Modelling can be an innovative technology or related to innovation [109, 110]. Modelling is preceded by correlation analysis [111, 112]. A prerequisite for modelling in space astronomy is the use of databases [113] and information systems [104]. One of the theoretical bases of information modelling in space astronomy is information theory [114].

In space astronomy, an information model is an abstract or formal construct that defines the dependence of changes in the state of the object of study. In space astronomy, an information model can be represented not only as a single model, but as a set of related and complementary models.

In space astronomy, information models can often describe certainty [115, 116] or uncertainty [117] depending on its parameters, input data, initial conditions and time. In analogue simulation, the model is built before the simulation process starts. It is a closed model. Information modelling allows refinement and modification of models, i.e. it is open-ended

An information model is a more flexible design than an analogue model. This is its advantage and versatility. In space astronomy, information models and information modelling reduce uncertainty. This is another advantage of information modelling.

In space astronomy, information uncertainty often serves as a sign of tacit knowledge [119- 121]. Hence, another task of information modelling in space astronomy is to extract implicit knowledge and transform it into explicit knowledge [122].

In space astronomy, information modelling is complemented by qualitative methods such as preference analysis [123]. In space astronomy, information modelling is complemented by semantic modelling [124]. Information uncertainty is also removed using cognitive methods [125-127].

Information models are a modern tool for building a picture of the world [128-131]. These models allow describing clear and fuzzy regularities and relationships in the information field. Currently, two types of mathematical modelling are widely used: analytical and simulation. Analytical modelling is possible only with known analytical dependencies describing the model and its relations with the external environment. It allows obtaining an exact solution based on mathematical descriptions linking objects and their parts.

The advantages of analytical modelling are objectivity, reproducibility, and communicability. If explicit dependencies linking the characteristics of the state of the object with initial conditions, parameters and variables of the system are known, then modelling gives correct results close to the real behaviour of the object. However, it is not always possible to obtain such dependencies. To use the analytical method it is often necessary to simplify the initial model significantly to be able to study the general properties of the system. The information model is less demanding. It allows the object to be described by fragments having different mathematical descriptions. A striking example of an effective "non-mathematical" model is the infological model [132]. In space astronomy, information modelling is not yet an established field. Many terms are borrowed from the theory of information modelling applied in terrestrial conditions. Therefore, there is a problem of terminological relations [133]. Therefore, there is a problem of what to apply from terrestrial informatics and what to create for space astronomy?

3. Gaining knowledge in space astronomy

Obtaining knowledge in space astronomy is based on the concept of information space field. The information space field is built as an integral model of reality. where all the information from outer space, received with the help of sensors, is downloaded.

The information space field is a "mapping" of the real space and all fields and objects that are in it. The basis of information space field research is knowledge acquisition. Fig.3.1 gives a scheme of formation of the information space field from the real space. Real fields contain objects that have integrity and completeness. Real objects have properties and have a life cycle of existence. Real objects can denote processes, phenomena, regularities. Real objects can be in groups or in situations. A group is a weakly connected aggregate that is united by common properties. A situation is a strongly connected aggregate, which is united by connections. Information field describes the content component of the real field. The concept of information field is an analogue of the concept of physical field

Fig.3.1 Transformation of reality into an information field.

The information field is an informational description of the real field and is formed as a model of transformation of the real field. the peculiarity of transformation of the real field into the information field is the indirectness of the transformation process. The information field is formed (Fig. 3.1) with the help of a device (measurement), with the help of a person (observation, reasoning), with the help of a model (modelling or using stereotypes). Real objects, as well as connections and relations between them, are transformed into information objects, information connections and information relations.

The concept of information objects is a generalised concept. It can denote: information models, information constructions, information situations, information units, information processes and information interactions [134]. Complementing information objects or filling the space between them, information relations and information links. information relations are also a generalised concept. They describe information relations in parameter space and describe spatial relations in coordinate space.

The information field reflects real spatial processes, regularities, real spatial and parametric interactions and real spatial relations. The information field is formed only when creating an information space that coordinates and systematises patterns.

An information field serves as an accumulator of real data and its subsequent transformation into information models and relations. An example of an information field is a satellite navigation field, in which three-dimensional coordinates are measured using satellite equipment. Geodata is a type of data and the result of its measurements in the navigation field.

The development of science and society requires periodic systematisation of knowledge. The formal basis of knowledge acquisition is information and intellectual resources [135]. The actual basis of knowledge retrieval is the information field. Knowledge retrieval can be related to Aristotle's definition of motion. He distinguishes six types of motion - emergence, annihilation, increase, decrease, transformation and movement. This can be related to knowledge extraction and manipulation. On the basis of extraction, knowledge emerges, new knowledge can lead to the destruction of old knowledge, knowledge can accumulate and increase, knowledge in a changing situation can decrease, knowledge can transform into new knowledge, knowledge can be transferred (moved). The diversity of definitions of "knowledge" and "information" (OTI) is due to the development of science and the application of these concepts in various spheres, which gives them a dynamic meaning.

Figure 3.2 shows the conceptual scheme of knowledge extraction. In any approach to knowledge retrieval, human involvement (natural intelligence of EI) is fundamental.

Figure 3.2. Knowledge extraction technologies

Real objects are immersed in the real field, which is located in real space. With the help of measurement technologies (I1), information is extracted from the real field and transferred to the information field. With the help of modelling technologies (M1), information models are formed from information (data). Models and information form information resources. These are not the only components of information resources. They include software, archives, libraries, digital models, maps, images and so on. In addition. in the information field there exist tacit knowledge that can be identified and form a special class of tacit knowledge models.

Fig. 3.2 identifies three fundamental approaches to knowledge acquisition. The first approach is based on the application of human intelligence (HI) only. It is applicable when information models as knowledge sources are observable and perceivable by a person (information barriers do not work), then knowledge is formed (obtained) by means of reasoning and logical methods (including unconventional ones). In this case, a person creates his/her own information models (e.g., infological models) to make inferences and obtain results. This approach also includes cognitive modelling using EI.

The second approach occurs when the initial data collections are so large as to preclude adequate human perception and analysis. The first approach becomes unacceptable. In this case, information technologies are applied that transform and simplify the original information in the form of secondary information models into a new form acceptable for human perception and analysis. This approach can be considered algorithmic, since information processing is carried out according to algorithms compiled by a person.

In this approach, information technology (IT) realises the technological aspect of information processing. In this case, the human being has the final say (IT +EI). Nevertheless, in this approach

the information load on a person is significantly reduced compared to the first approach. The third approach is used when the first and the second are unacceptable. It occurs when the initial information is not only large in volume for human perception, but is so complex that it cannot be processed using simple algorithms and information technologies and systems. In this case, artificial intelligence (AI) methods are resorted to. This approach is called intelligent approach [139-139].

In this approach, intelligent technologies implement not only the technological aspect of information processing, but also the cognitive aspect of its analysis after processing and knowledge acquisition (AI+EI). In this approach, the information load on a human being is significantly reduced compared to the second approach. In this approach, intellectual technologies realise the extraction of implicit knowledge and its transformation into explicit knowledge.

As a result of data collection and information accumulation, information models are formed, which are the result of reflection or information morphism of information about real objects

Information model - identifiable and recognisable object with data and/ or prescriptions concerning data, stable in time and limited in virtual information space. Description of some entity (real object, phenomenon, process, event) in the form of a set of logically connected requisites - information elements. Models are used to describe and study objects. Reflection of a real object is an information model.

Information model is a purposeful formalised representation of an existing object of reality or a set of objects by means of a system of interrelated, identifiable, informatively defined parameters. Information model provides formalised representation of used data and their interrelationships

As a result of collection in information technologies, primary (raw) data in different formats are obtained. For the possibility of sharing and comparability of heterogeneous data, data unification procedures are applied (Fig. 3.3), which transform different forms of representation and formats into a single data system. Data harmonisation is a mandatory procedure in modern complex technologies of data collection using different technologies and different instruments.

Fig.3.3 Transformation of initial data.

As a result of unification, a complementary data system is formed, including primary information models - data models. These models can be simple or composite, but all of them are brought to a form convenient for their joint processing.

After data unification, information (data) models containing the description of the object and its microenvironment are obtained. These information models are used for processing in information systems. These models are processed in information technology (IT) and information systems (IS). In geoinformatics, they are processed in GIS.

As a result of information processing in information technologies, information systems or GIS, secondary information models are obtained, which, unlike primary models, contain explicit or implicit knowledge.

Figure 3.4. shows the process of knowledge acquisition based on the interrelationship information - patterns - knowledge.

Fig.3.4 Information, models, knowledge

Models are used for storage or for application tasks. They can be used for commercial purposes. Information technology and information systems serve as technological tools for acquiring new knowledge. An additional tool for extracting new knowledge is human intelligence. This is due to the fact that information systems represent human-machine systems.

Information morphism and metamodelling are applied to extract tacit knowledge. Morphism as a transformation between real and information field is one of the types of description of information processes. including transformation. It gives a description at the level of categories and meta-models. Morphism preserves the structure of the original object while transferring it into an information model. An example is a topographic map, which as a model preserves topological spatial relations of reality. Morphism can be considered as a transformation within an information field. The mechanism of morphism description allows to perform a generalised analysis of technologies and processes in the information field. Morphism in the information field allows generalising and identifying concealment and latent processes and facilitating the extraction of tacit knowledge.

The information field is a representation of the real field. For this mapping it is necessary to create an information space beforehand. To create an information space, a preliminary formalism of its description is necessary. For example, a spatial coordinate space can be described in a Cartesian coordinate system, a cylindrical coordinate system or a spherical coordinate system. The choice of the coordinate system is dictated by the problem to be solved and the type of the real field. Information processes, categorical processes and meta-processes occur in the information field. This requires the involvement of category theory and metamodelling for the analysis of the field. Knowledge extraction is carried out from the information field, which includes primary and secondary data, primary and secondary models. On the basis of modelling and metamodelling, new knowledge is identified. Including regularities and laws of the real field through the information field.

Conclusion

The monograph explores space astronomy as a new scientific direction. Space astronomy complements classical astronomy and is its development. Space astronomy can be considered as the astronomy of points of observation of which observations are transferred from the Earth's surface and near-Earth space into space on space carriers. Space astronomy differs from classical astronomy not only in the transfer of observation points into outer space, but also in its methodological and organisational basis. Space astronomy, in addition to the basics of astronomy, uses as components new sciences: space geodesy, space geoinformatics, geodesic astronomy, spatial analysis, spatial logic, systems analysis, informatics. The application of spatial logic is one of the features of space astronomy. The main means of observation in space astronomy is space monitoring, which uses the ideas of geo-information monitoring. Space astronomy is a "situational" science. It not only observes, but also measures and forms management decisions. For example, the task of selecting landing sites for spacecraft belongs to the field of space astronomy. The application of the information field model is an important distinction of space astronomy. The information field is an integral model that combines different fields and allows for complex analyses and complex problem solving.

Literature

1   Zombeck, M. V. (2006). Handbook of space astronomy and astrophysics. Cambridge University Press

2   Gospodinov S. G. The Development of Geodesic Astronomy // Russian Journal of Astrophysical Research. Series A, 2018, 4(1). C. 9-33

3   Gospodinov S.G. Evolution of geodetic astronomy // Russian Journal of Astrophysical Research. Series A, 2022, 8(1). C. 3-11.

4   .Bondur V. G., Tsvetkov V. Ya.. New Scientific Direction of Space Geoinformatics // *European Journal of Technology and Design,* 2015. 4(10), 118126/

5.   Savinykh V.P. Development of space geoinformatics // Slavic Forum, 2016. - 2(12). - c.223-230

6.   Tsvetkov V.Y., Savinykh V.P. Space geoinformatics: textbook for universities. - St. Petersburg: Lan, 2022. - 184 c.

7.   Savinykh V.P. State of Space Geoinformatics. Monographic article // Slavic Forum. 2021, 2(32). C.7-17

8.   Raev V.K. Dynamic geoinformatics // Slavic Forum. 2022, 2(36). C. 195-2053

9.   Tsvetkov V.Ya. Spatial logic in education and science // Educational resources and technologies. - 2019. - № 2 (27). - C. 92-102.

10.   Tsvetkov V.Ya. Spatial logic in geoinformatics // Vector GeoSciences. 2020. T. 3. № 2. C. 91-100.

11.   Tyagunov A.M., Tsvetkov V.Ya Logic of Space Observations // Russian Journal of Astrophysical Research. Series A, 2021,7(1): 43-48

12.   Bondur V. G., Tsvetkov V. Ya. System Analysis in Space Research // Russian Journal of Astrophysical Research. Series A. 2015. №1(1), c. 4-12

13.   Polyakov A.A., Tsvetkov V.Y. Applied Informatics. - Moscow: Janus-K, 2002. - 392 c.

14.   Gospodinov S.G. Modern Informatics. - Saarbruken, 2023. - 157 c.

15.   Oznamets V.V. The Evolution of Space Geodesy // Russian Journal of Astrophysical Research. Series A, 2023, 9(1). C. 14-19

16.   Oznamets B.B., Tsvetkov V.Ya. Space geodesy of small celestial bodies // Russian Journal of Astrophysical Research. Series A. 2019, 5(1). c. 30-40.

17.   Kudzh S.A. Development of space monitoring // Russian Journal of Astrophysical Research. Series A, 2022,8(1). C.12-22

18.   Tsvetkov V.Ya. Space monitoring: Monograph. - Moscow: MAKS Press, 2015. - 68 c.

19.   Savinych V. P. Evolution of Space Monitoring // Russian Journal of Astrophysical Research. Series A. 2017. -3(1). c. 33-40

20.   Monakhov C.B., Savinykh V.P., Tsvetkov V.Y. Methodology of Analysis and Design of Complex Information Systems. - Moscow: Prosveshchenie, 2005. - 264 c.

21.   Raev V.K. Information space and information field // Slavic Forum. 2021, 4(34). C. 87-96

22.   Bondur V.G. Information fields in space research // Educational resources and

technologies. - 2015. - №2 (10). - c.107-113.

23.   Gospodinov S.G. Information Fields in Space Research // Russian Journal of Astrophysical Research. Series A, 2023,9(1). C. 10-13.

24.   Kudzh S. F., Tsvetkov V.Ya. Spatial logic concepts. *Revista inclusions.* 2020. Volumen 7. Numero Especial / Julio - Septiembre. 837-849

25.   Savinykh V.P. Determination of the linear parameters of the planet from the angular diameter measurement // Russian Journal of Astrophysical Research. Series A, 2021,7(1): 28-34

26.   Tsvetkov V.Ya. Angular measurements in space geoinformatics // Russian Journal of Astrophysical Research. Series A, 2021,7(1): 35-42.

27.   Rosenberg I.N., Tsvetkov V.Y. Coordinate systems in the Geoinformatics - MSUPS, 2009. -67 c.

28.   Maksudova L.G., Savinykh V.P., Tsvetkov V.Ya. Integration of sciences about the surrounding world in geoinformatics // Earth Exploration from Space. - 2000. - №1. - c.46-50

29.   Ivannikov A.D., Kulagin V.P., Tikhonov A.N., Tsvetkov V.Y. Applied Geoinformatics. - Moscow: MAKS Press, 2005. -360 c.

30.   Savinykh V.P., Tsvetkov V.Ya. Development of artificial intelligence methods in geoinformatics // Transport of the Russian Federation. - 2010. - № 5. - c.41-43

31.   Tsvetkov V.Ya. Information model as a basis for information processing in GIS // Izvestiya vysokikh uchebnykh uchebnykh obrazovaniye. Geodesy and aerial photography. -2005. - №2. - c.118-122.

32.   Tsvetkov V.Ya. Digital maps and digital models // Izvestiya vysshee obrazovaniye vysshee obrazovaniye. Geodesy and aerial photography. - 2000. - №2. - c.147-155.

33.   Tsvetkov, V.Ya. The use of oppositional variables to analyse the quality of educational services (in Russian) // Sovremennye naukoyemkie tekhnologii. - 2008. - №1. - c.62-64

34.   Tsvetkov V.Ya. Geoinformation monitoring // Izvestiya vysshee obrazovaniya vysshee obrazovaniya. Geodesy and aerial photography. - 2005. - №5. - c.151 -155

35.   Tsvetkov V.Ya., Shaitura S.V., Minitaeva A.M., Feoktistova V.M., Kozhaev Yu.P., Belyu L.P. Metamodelling in the information field // Amazonia Investiga. 2020. T. 9. № 25. C. 395-402

36.   Tsvetkov V.Ya. Information field and information space // International Journal of Applied and Fundamental Research. - 2016. - №1-3. - c.455-456.

37.   Hariri R. H., Fredericks E. M., Bowers K. M. Uncertainty in big data analytics: surveys, opportunities, and challenges // Journal of Big Data. - 2019. - T. 6. - №. 1. - C. 1-16.

38.   Buravtsev, A.V.; Tsvetkov, V.Ya. Cloud computing for big geospatial data // Information and Space. 2019. - №3. -p .IO- 115.

39.   Levin B.A., Tsvetkov V.Ya. Information processes in the space of "big data" // World of Transport. 2017. - T.15, №6(73). - c.20-30.

40. Reddy G. T. et al. Analysis of dimensionality reduction techniques on big data //Ieee Access. - 2020. - T. 8. - C. 54776-54788.

41. Tsvetkov V.Ya. Formation of spatial knowledge: Monograph. - Moscow: MAKS Press, 2015. - 68 c.

42. Savinych V. P. On the Relation of the Concepts of Space Knowledge, Knowledge, Knowledge of the Spatial // Russian Journal of Astrophysical Research. Series A. 2016. 1(2), p. 23-32.

43. 43. Gospodinov S.G., Tsvetkov V.Y. Some theoretical aspects of instrumental methods of space astronomy. Sofia. Art Eternal Cinema. 2023. - 92 c.

44. Tsvetkov V.Ya. Globalisation and Informatisation // Information Technologies. - 2005. - №2. - c.2-4

45. Savinykh V. P., Tsvetkov V. Ya. Geoinformatics as a system of sciences // Geodesy and Cartography. - 2013. - №4. - c.52-57

46. Omelchenko A. C. Geodata as an innovative resource // Quality, Innovations, Education. - 2006. - №1. - c.12- 14.

47. Kuj A.S. Collection and measurement of geodata in Earth Sciences// Slavic Forum. - 2013. - 2(4). - c.135-139

48. Savinykh V.P., Tsvetkov V.Y. Geodata as a system information resource // Bulletin of the Russian Academy of Sciences, 2014, Vol. 84, No. 9, pp. 826-829. DOI: 10.7868/S0869587314090278

49. Kovalenko N.I. Spatial reasoning with application of geodata// Slavic Forum. -2020. - 4(30). -c.273-283.

50. Tsvetkov V.Ya. Types of spatial relations // Uspekhi sovremennoi naukhnostvosnaniya. - 2013. - № 5. - c. 138-140.

51. Tsvetkov V.Ya. On spatial and economic relations // International Journal of Experimental Education. - 2013. - №3. - c.115-117.

52. Savinykh V.P. Information spatial relations // Educational resources and technologies. - 2017. - №1 (18). - c. 79-88.

53. Kovalenko A.N. System approach of creating an integrated information model // Slavic Forum. - 2014. - 2 (6). - c.51 -55

54. Tsvetkov V.Ya. Creation of integrated information basis of GIS// Izvestiya vysshee obrazovaniya vysshee obrazovaniya [Izvestia of Higher Educational Institutions. Geodesy and aerial photography. - 2000. - №4. - c.150-154.

55. Tsvetkov V.Ya. Geostatistics // Izvestiya vysokikh uchebnykh obrazovaniye. Geodesy and aerial photography. - 2007. - №3. - c.174-184.

56. Tsvetkov V.Ya., Zaitseva O.V. Geostatistics as a management tool // Izvestiya vysokikh uchebnykh obrazovaniya. Geodesy and aerial photography. - 2007. - №5. - c. 134 - 137.

57. Deshko I.P. Information approach in modelling // Educational resources and technologies. - 2016. - №5 (17). - c.21-26.

58. Rozenberg I.N., Tsvetkov V.Ya. The Geoinformation approach // Eurupean Journal of Natural History. - 2009. - №5. - p. 102 -103.

59.   Zatyagalova B.B. Geoinformation approach in monitoring of sea pollution from remote sensing data from space // Earth Sciences. - 2-2012.- c.80-85.

60.   Tsvetkov V.Ya. Visual modelling in decision support systems // International Journal of Applied and Fundamental Research. - 2016. - №10-1. - c.13-17.

61.   Nepoklonov V.B., Shlapak V.V., Tsvetkov V.Ya., Lonskiy I.I. Automation of visual geospatial information processing
GIS // Information and Space. -2018. - №2. - c. 93-97

62.   Tsvetkov V.Ya. Relation, connection, correspondence // Slavic Forum, 2016. -2(12). - c.272-276

63.   Ozherel'eva T.A. On the relation between the concepts of information space, information field, information environment and semantic environment // International Journal of Applied and Fundamental Research. - 2014. - № 10 - c. 21-24

64.   Tsvetkov V. Ya. Information Space, Information Field, Information Environment // European researcher. 2014. № 8-1(80). p.1416-1422.

65.   Tsvetkov V. Ya. Natural and artificial information field// International Journal of Applied and Fundamental Research. -2014. - №5-2. - c.178 -180.

66.   Savinykh V.P., Tsvetkov V.Ya. Geoinformation analysis of remote sensing data. - Moscow: Kartocentre-Geodesizdat, 2001. - 224 c.

67.   Chekharin E. E. Algorithms of remote sensing data interpretation // Slavic Forum. - 2015. - №. 3. - C. 301-308.

68.   Savinykh V.P., Tsvetkov V.Ya. Features of integration of geoinformation technologies and remote sensing data processing technologies // Information Technologies. - 1999. - №10. - c.36-40.

69.   Savinykh V.P., Tsvetkov V.Ya. Integration of GIS technologies and Earth remote sensing systems // Earth Exploration from Space.- 2000. - №2 - c.83-86.

70.   Tsvetkov V.Ya. Diversification of space monitoring // Slavic Forum, 2015. - 2(8) - c.302-309.

71.   Hayford J. F. A text-book of geodetic astronomy. - J. Wiley, 1910.

72.   Ramsayer I. K. Handbuch der Vermessungskunde: Band 2A: Geodatische Astronomie. - Springer-Verlag, 2017.

73. Gospodinov G.S. Geodesic astronomy and space geoinformatics // Railways Science and Technology. - 2017. T.1. - 1(1). - c.45-50.

74.   Slaveiko Gospodinov, Severina Jordova. Geodesic astronomy.- Military Geographical Service (Bulgaria), 2011. -264c.

75.   Kraus J. D. Radio astronomy - New York: McGraw-Hill, 1966.

76.   Robbins A. R. Time in geodetic astronomy //Survey Review. - 1967. - V. 19. - №. 143. - p.2-19.

77.   I.V. Barmin, V.P. Kulagin, V.P. Savinykh, V.Ya. Tsvetkov. Near_Earth Space as an Object of Global Monitoring // Solar System Research, 2014, Vol. 48, No. 7, pp. 531-535. DOI: 10.1134/S003809461407003X.

78.   Savinykh V.P., Tsvetkov V.Ya. Comparative planetology. - Moscow: MIIGAiK, 2012, -84 pp.

79.   Kulagin V.P., Savinykh V.P., Tsvetkov V.Ya. Features of studying the discipline "Comparative planetology"// Perspectives of Science and Education. - 2013. -№5. - C.129-133.

80.   V. Ya. Tsvetkov. The Development of the Direction "Comparative Planetology" // Russian Journal of Astrophysical Research. Series A, 2018, 4(1): 34-41.

81.   Taylor F. W. Comparative planetology, climatology and biology of Venus, Earth and Mars //Planetary and Space Science. - 2011. - V. 59. - №. 10. - p. 889-899.

82.   Matson D. L. et al. Asteroids and comparative planetology //Lunar and Planetary Science Conference Proceedings. - 1976. - T. 7. - C. 3603-3627.

83.   Bean J. L., Abbot D. S., Kempton E. M. R. A statistical comparative planetology approach to the hunt for habitable exoplanets and life beyond the solar system //The Astrophysical Journal Letters. - 2017. - T. 841. - №. 2. - C. L24.

84.   Kuchner M. J. General astrophysics and comparative planetology with the Terrestrial Planet Finder missions //arXiv preprint astro-ph/0505102. - 2005/

85.   V. Ya. Tsvetkov. The Logarithmic Measure the Orbits of the Planets of the Solar System // Russian Journal of Astrophysical Research. Series A. 2017. - 3(1). pp. 41-46/

86.   Savinych V.P. Study of the "earth-moon" system // Russian Journal of Astrophysical Research. Series A, 2022,8(1). C. 32-39

87.   Tikhonov V. E., Neverov A. A. The Earth's motion around the solar system barycentre as an information basis for long-term forecasting of crop yield // Agrarny Vestnik Urala. - 2015. - №. 12 (142).

88.   Jantsch P., Webster C. G., Zhang G. On the Lebesgue constant of weighted Leja points for Lagrange interpolation on unbounded domains //IMA Journal of Numerical Analysis. - 2019. - T. 39. - №. 2. - C. 1039-1057.

89.   Capdevila L. R., Howell K. C. A transfer network linking Earth, Moon, and the triangular libration point regions in the Earth-Moon system //Advances in Space Research. - 2018. - T. 62. - №. 7. - C. 1826-1852.

90.   Tsvetkov V.Y., Savinykh V.P. Space geoinformatics: textbook for universities. - St. Petersburg: Lan, 2022. - 184 c.

91.   Zimovan E. M. Characteristics and design strategies for near rectilinear halo orbits within the Earth-Moon system : doi. - Purdue University, 2017.

92.   Tsvetkov V.Ya. Information-measuring systems and technologies in geoinformatics. - Moscow: MAKS Press, 2016. - 94c.

93.   Ozherelieva T.A. Information modelling in educational technologies // International Journal of Applied and Fundamental Research. - 2016. - №3-2. - c.214-218

94.   Tsvetkov V.Ya. Information models and information resources // Izvestiya vysshee obrazovaniya vysshee obrazovaniya [Izvestia of higher educational institutions]. Geodesy and aerial photography. - 2005. - №3. - c.85-91

95.   Maksudova L.G., Tsvetkov V.Ya. Information modelling as a fundamental method of cognition // Izvestiya vysokikh uchebnykh obrazovaniya [Izvestia of Higher

Educational Institutions].
Geodesy and aerial photography. - 2001. - №1. - c.102-106.
96.   Bolbakov R. G. Semiotic information modelling // Slavic Forum, 2015. - 4(10) - c.54-60
97.   Deshko, I. P.; Matchin, V. T. Geoinformation compositional modelling // Information and Space. 2022. - №2. -c .180-184.
98.   Tsvetkov V.Ya. Fundamentals of geoinformation modelling // Izvestiya vysokikh uchebnykh obrazovaniya. Geodesy and aerial photography. - 1999. - №4. - c.147 -157.
99.   Andreeva O.A. Geoinformation modelling // Slavic Forum. -2019. - 2(24). - c.7-12.
100.   Tsvetkov V.Ya. Modelling of scientific research in automation and design. - Moscow: SCST, VNTIcentre, 1991. - 125c.
101.   Tsvetkov V.Ya. Information description of the world picture // Prospects of science and education. - 2014. - №5(11). - c.9-13
102.   Tsvetkov V.Ya. Knowledge extraction for the formation of information resources. - M.: GNII IOT. 2006. - 158 c.
103.   Savinych V.P. Digital Simulation in Space Research // Russian Journal of Astrophysical Research. Series A, 2023,9(1). C. 20-23.
104.   Savinykh V.P. Information systems for planetary studies // Russian Journal of Astrophysical Research. Series A. 2019, 5(1). C.41-55.
105.   Shaitura C.B. Modelling and construction // Slavic Forum. -2019. - 1(23). - c.68-79
106.   Rogov I.E. Information Modelling. - Saarbruken, LAP Lambert Academic Publishing, 2020. -105 c.
107.   Deshko I.P. Communication closed information model // Educational resources and technologies - 2017. -3 (20). - c.41-47.
108.   Anikina G.A., Polyakov M.G., Romanov L.N., Tsvetkov V.Ya. On image contour extraction using linear trained models.
// Izvestia of the USSR Academy of Sciences. Technical Cybernetics. -1980. - №6. - c.36-43
109.   Tsvetkov V. Ya. Informatisation, innovation processes and geoinformation technologies // Izvestiya vysokikh uchebnykh obrazovaniya. Geodesy and aerial photography - 2006. - №4. - c.112-118/
110.   Rosenberg I.N., Soloviev I.V., Tsvetkov V.Ya. Complex innovations in the management of complex organisational and technical systems. / edited by V.I. Yakunin - M.: Feoria, 2010. - 248 c.
111.   Tsvetkov V.Ya., Obolyayeva N.M. Using correlative approach for personnel management of educational institution // Distance and virtual learning. - 2011. - №8. - c.4- 9.
112.   Tsvetkov V. Ya. Framework of Correlative Analysis // European researcher.2012. № 6-1 (23). C. 839-844

113.   Tsvetkov V.Ya. Designing data structures and databases - M.: Moscow State University of Geodesy and Cartography, 1997. - 90 c.

114.   Ivannikov A.D., Tikhonov A.N., Tsvetkov V.Ya. Fundamentals of information theory - M.: MAKS Press, 2007. - 356 c

115.   Ozherel'eva T.A. Oppositional analysis of uncertainty and certainty // Slavic Forum. - 2017. - 1(15). - c.218-226.

116.   Tsvetkov V.Ya. Information uncertainty and certainty in information sciences // Information Technologies. - 2015. - №1. -c.37

117.   Nomokonova O. Yu. Information uncertainty in information interaction // Slavic Forum. - 2017. -1(15). - c.104-110.

118.   Elsukov P.Y. Information reducing uncertainty and information increasing meaningfulness // Educational Resources and Technologies - 2017. -3 (20). - c.62-68.

119.   Tsvetkov V. *Ya.* Analysis of implicit knowledge // Perspectives of Science and Education. - 2014. - №1 (7). - c.56-60

120.   Tsvetkov V.Ya. Implicit knowledge and its varieties // Vestnik of Mordovian University. - 2014. - T. 24. № 3. - c.199-205.

121.   Kuj S. A. Implicit knowledge in the information field // Slavic Forum. -2018. - 3(21). - c. 14-20.

122.   Savinykh V.P. Explicit and implicit knowledge // Slavic Forum. -2020. - 2(28). -c.103-111.

123.   Tsvetkov V. Ya. Not Transitive Method Preferences. // Journal of International Network Centre for Fundamental and Applied Research. 2015. 1(3), - pp.34-42. DOI: 10.

124.   Tsvetkov V. Ya. The Semantic environment of information units // European researcher. 2014, № 6-1 (76). p. 1059-1065

125.   Elsukov P. Yu. Cognitive logic // Slavic Forum. -2020. - 3(29). -c.87-95.

126.   Tsvetkov V.Ya. Cognitive aspects of building virtual educational models// Perspectives of Science and Education. - 2013. -№3 (3). - c.38-46

127.   Rogov I.E., Chekharin E.E. Cognitive constructions of information search. // Slavic Forum. -2020. - 3(29). -c.150-159.

128.   Tsvetkov V. Ya. Formation of the world picture // European Journal of Technology and Design, 2020, 8(1). 33-37.

129.   Butko E.Ya. Personal picture of the world as a result of education // Distance and virtual learning. 2017. - № 1 (115). - c.87-94.

130.   Chekharin E. E. The world picture as a cognitive model // Slavic Forum, 2016. -4(14). - c.290-296.

131.   Butko E. Я. Geoinformatics as a method of building a picture of the world // Slavic Forum. - 2017. -1(15). - c.34-41.

132.   Raev V.K. Infological models as a research tool //
Slavic Forum. -2020. - 3(29). -c.56-66.

133.   Tikhonov A.N., Ivannikov A.D., Tsvetkov V.Y. Terminological relations // Fundamental Research. -2009. - № 5. - c.146- 148.

134.   Kuj S. A. Information interaction and its attributes// Slavic Forum. - 2017. - 4(18). - c.27-33.

135.   Zhang H. et al. Intelligent resources management system design in information centric networking //China communications. - 2017. - T. 14. - №. 8. - C. 105-123.

136.   Tsvetkov V.Y., Rosenberg IN. Intelligent transport systems - Saarbrucken, 2012. - 297 c.

137.   Rosenberg I.N., Tsvetkov V.Ya. Application of multi-agent systems in intellectual logistics systems. // International Journal of Experimental Education. - 2012. - №6. - c.107-109

138.   Buchkin D.V. State and development of intellectual GIS // Information and Space. 2020. - №3. -c .119-123.

139.   Kuzhelev, P.D. Intellectual multi-purpose management // State Counsellor. - 2014. - №4. - c.65-68

yes I want morebooks!

Buy your books fast and straightforward online - at one of world's fastest growing online book stores! Environmentally sound due to Print-on-Demand technologies.

Buy your books online at
www.morebooks.shop

Kaufen Sie Ihre Bücher schnell und unkompliziert online – auf einer der am schnellsten wachsenden Buchhandelsplattformen weltweit! Dank Print-On-Demand umwelt- und ressourcenschonend produziert.

Bücher schneller online kaufen
www.morebooks.shop

Printed by Books on Demand GmbH, Norderstedt / Germany